SpringerBriefs in

MW01519717

Green Chemistry for Sustainability

Series editor

Sanjay K. Sharma, Jaipur, India

More information about this series at http://www.springer.com/series/10045

Yebo Li · Xiaolan Luo · Shengjun Hu

Bio-based Polyols
and Polyurethanes

 Springer

Yebo Li
Department of Food, Agricultural and
   Biological Engineering
The Ohio State University, Ohio
   Agricultural Research
   and Development Center
Wooster, OH
USA

Shengjun Hu
Department of Food, Agricultural and
   Biological Engineering
The Ohio State University, Ohio
   Agricultural Research and Development
   Center
Wooster, OH
USA

Xiaolan Luo
Department of Food, Agricultural and
   Biological Engineering
The Ohio State University, Ohio
   Agricultural Research and Development
   Center
Wooster, OH
USA

ISSN 2191-5407   ISSN 2191-5415   (electronic)
SpringerBriefs in Molecular Science
ISSN 2212-9898
SpringerBriefs in Green Chemistry for Sustainability
ISBN 978-3-319-21538-9   ISBN 978-3-319-21539-6   (eBook)
DOI 10.1007/978-3-319-21539-6

Library of Congress Control Number: 2015945330

Springer Cham Heidelberg New York Dordrecht London

Springer International Publishing AG Switzerland is part of Springer Science+Business Media
(www.springer.com)

Preface

Polyurethanes are versatile polymeric materials and are usually synthesized by isocyanate reactions with polyols. Due to the variety of isocyanates and polyols, particularly polyols, polyurethanes can be easily tailored for wide applications, such as rigid and flexible foams, coatings, adhesives, and elastomers. Considerable efforts have been recently devoted to developing bio-based substitutes for petroleum-based polyurethanes due to increasing concerns over the depletion of petroleum resources, environment, and sustainability. This book first introduces general production methods and characteristics of polyols, isocyanates, and polyurethanes (Chap. 1) and then focuses on the synthesis and properties of bio-based polyols and polyurethanes from different renewable feedstocks including vegetable oils and their derivatives (Chap. 2), lignocellulosic biomass (Chap. 3), and protein-based feedstocks (Chap. 4). A comparison of bio-based polyol and polyurethane properties with their petroleum-based analogues is also discussed. This book provides useful information on bio-based polyols and isocyanates and their derived polyurethanes for chemists, researchers, students, and others who are interested in green chemistry and bio-based products.

Wooster, OH, USA

Yebo Li
Xiaolan Luo
Shengjun Hu

Acknowledgments

We would like to thank Mrs. Mary Wicks (Department of Food, Agricultural and Biological Engineering, Ohio State University) and Dr. Longhe Zhang (Case Western Reserve University) for reading through the book manuscript and providing useful suggestions. We also thank the series editor, Prof. Sanjay K. Sharma; the publishing editor, Dr. Sonia Ojo, and her team member, Esther Rentmeester; and others from Springer for their support, patience, and interest in this book.

Contents

Chapter 1
Introduction to Bio-based Polyols and Polyurethanes

Abstract Polyurethanes (PUs) are one of the most versatile polymers and are widely used in our daily lives for rigid and flexible foams, coatings, films, and other products. PUs are generally synthesized through reactions between isocyanates and polyols. A brief overview of the chemical structures, origin, synthetic methods, and properties of polyols and isocyanates is given in this chapter. Currently, most polyols are petroleum-based, but increasing concerns over the depletion of petroleum resources, environment, and sustainability have led to considerable efforts to develop bio-based polyols and PUs from renewable resources. Bio-based polyols and isocyanates for the production of bio-based PUs are discussed in this chapter.

Keywords Overview · Polyols · Polyurethanes · Renewable · Vegetable oils · Fatty acid derivatives · Biomass · Isocyanates

1.1 Introduction

Since the pioneering work on polyurethanes (PUs) was conducted by Bayer and his coworkers in 1937 [1], a large variety of PU products with versatile properties have been produced and extensively used as rigid and flexible foams, elastomers, adhesives, coatings, resins, and other uses. Among these PU applications, rigid and flexible foams are predominant and primarily used in construction, transportation, bedding and furniture, and packaging industries. The PU foam market was about $40.1 billion in 2012 and is estimated to reach $61.9 billion in revenue by 2018 [2].

Although PUs can be produced with and without the involvement of isocyanates, the reactions between polyols and isocyanates are most widely used, especially for commercial PU production. For PUs synthesized without isocyanates, various pathways are involved, including the reaction between cyclic carbonates and polyfunctional amines [3, 4], self-condensation of AB-type monomers containing hydroxyl and acyl azide groups [5], transurethane reactions between diurethanes and diols [6], and reactions between AB-type monomers containing hydroxyl and methyl carbamate groups [5]. In this book, we focus on the production of PUs through reactions of polyols with isocyanates.

© The Author(s) 2015
Y. Li et al., *Bio-based Polyols and Polyurethanes*,
SpringerBriefs in Green Chemistry for Sustainability,
DOI 10.1007/978-3-319-21539-6_1

1.2 Polyols

Polyols generally refer to compounds that contain multiple reactive hydroxyl groups in one molecule. Depending on molecular weight, polyols can be classified into two categories, i.e., monomeric and polymeric polyols [7]. Monomeric polyols are low molecular weight organic compounds, such as glycerol, ethylene glycol (EG), propylene glycol, diethylene glycol (DEG), and 1,4-butanediol (BD). According to the number of hydroxyl groups in one molecule, monomeric polyols are also classified as diols which have two hydroxyl groups in one molecule, triols which have three hydroxyl groups in one molecule, tetraols which have four hydroxyl groups in one molecule, etc. Usually, diols are useful chain extenders in the production of PUs. Monomeric polyols with three or more hydroxyl groups in one molecule, such as triols and tetraols, are widely used as initiators for the production of polyether polyols or crosslinkers for the fabrication of PUs.

Polymeric polyols are high molecular weight polymers, which are major components for the formation of PU backbones. Polyether and polyester polyols are the two most common polymeric polyols. Petroleum-based polyether polyols are commonly produced through alkoxylation of alkylene oxide monomers, such as ethylene oxide, propylene oxide, or mixtures of these two monomers, under base catalysis with a multi-hydroxyl alcohol (e.g., glycerol, pentaerythritol) as an initiator. Base catalysts such as alkali metal hydroxides are generally used in the alkoxylation process [8]. The alkoxylation reaction of propylene oxide for the production of polyether polyols using glycerol as an initiator is shown in Scheme 1.1. The produced poly(propylene oxide) polyols have secondary terminal hydroxyl groups, which are less reactive than primary hydroxyl groups. To improve the reactivity of poly(propylene oxide) polyols, ethylene oxide is added in the final stage of the alkoxylation reaction to introduce primary hydroxyl end groups [9]. The introduction of ethylene oxide can also improve the hydrophilicity of polyols. The functionality of polyether polyols largely depend on the functionality of the initiator used in the alkoxylation process. For example, poly(propylene oxide) polyols produced with glycerol as an initiator have a functionality of three (Scheme 1.1). By controlling the functionality of the initiator and the degree of polymerization, a large range of polyether polyols with different functionalities and

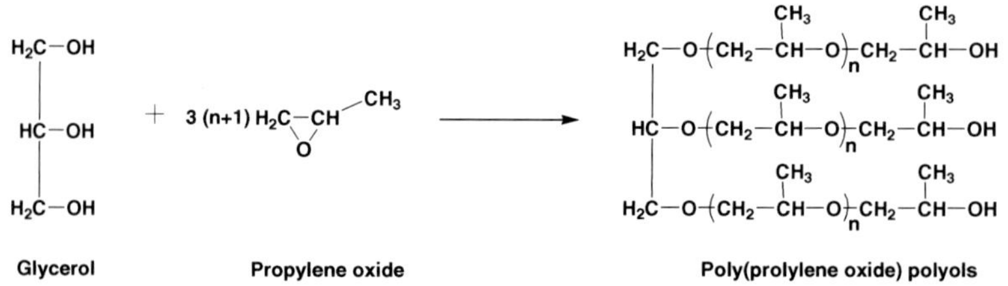

Scheme 1.1 Production of poly(prolylene oxide) polyols via alkoxylation of propylene oxide using glycerol as an initiator

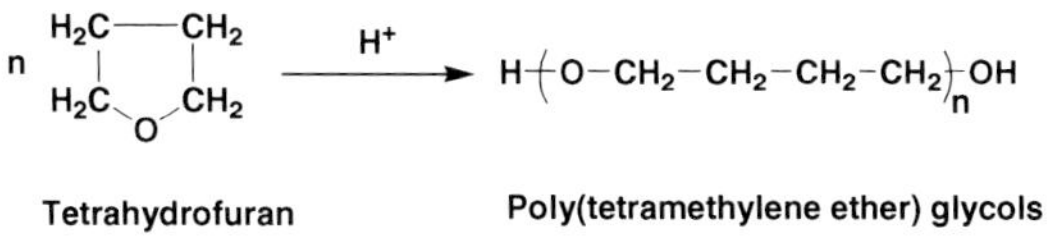

Scheme 1.2 Production of poly(tetramethylene ether) glycols via acid-catalyzed polymerization of tetrahydrofuran

molecular weights can be obtained for PU applications. Generally, polyether polyols with high functionality (3.0–8.0), low molecular weight (150–1000 g/mol), and high hydroxyl number (250–1000 mg KOH/g) are suitable for the production of rigid foams, rigid coatings, and elastoplastics, whereas polyether polyols with low functionality (2.0–3.0), high molecular weight (1000–6500 g/mol), and low hydroxyl number (28–160 mg KOH/g) are used for the production of flexible foams and elastomers [8, 9]. Compared to polyester polyols, alkoxylation-derived polyether polyols are low viscosity liquids and have superior hydrolytic stability and chemical resistance. Another type of important polyether polyols, poly(tetramethylene ether) glycols, are produced by acid-catalyzed polymerization (Scheme 1.2). Poly(tetramethylene ether) glycols are usually waxy solids at room temperature, and melt into liquids when heated to about 40 °C [10]. Their major applications include thermoplastic PUs, PU elastomers, and PU fibers.

Polyester polyols with linear structures are mainly produced by condensation polymerization of dicarboxylic acids or their esters or anhydrides with monomeric diols, such as diethylene glycol and 1,4-butanediol. With the addition of triols, such as glycerol and 1,1,1-trimethylol propane, branched polyester polyols can be obtained. Based on the polyol structure, polyester polyols can be classified into aliphatic and aromatic polyester polyols. Adipic acid and phthalic anhydride are the most commonly used dicarboxylic acid/anhydride in the production of aliphatic and aromatic polyester polyols, respectively. Typical polyester polyols obtained with the reaction between adipic acid and 1,4-butanediol, or between phthalic anhydride and diethylene glycol, are shown in Scheme 1.3. Aliphatic polyester polyols with low hydroxyl numbers are usually used for the production of PU elastomers, adhesives, and some flexible foams. Compared to aliphatic polyester polyols, aromatic ones provide higher rigidity and are commonly used for the production of flame-retardant rigid PU foams. Polyester polyols are usually waxy solids or liquids with high viscosity. As ester groups are more susceptible to hydrolytic attack than ether groups, polyester polyols are less water-resistant than polyether polyols. Polycaprolactone polyols, which are another type of aliphatic polyester polyol, are synthesized by ring-opening polymerization of ε-caprolactone using various multi-hydroxyl alcohols as initiators (e.g., 1,4-butanediol, Scheme 1.4). They have better water resistance and lower viscosity than adipic acid-based polyester polyols, which have similar chemical structures and comparable molecular weights [9]. Polycaprolactone polyols have broad applications in foams, elastomers, coatings, and adhesives.

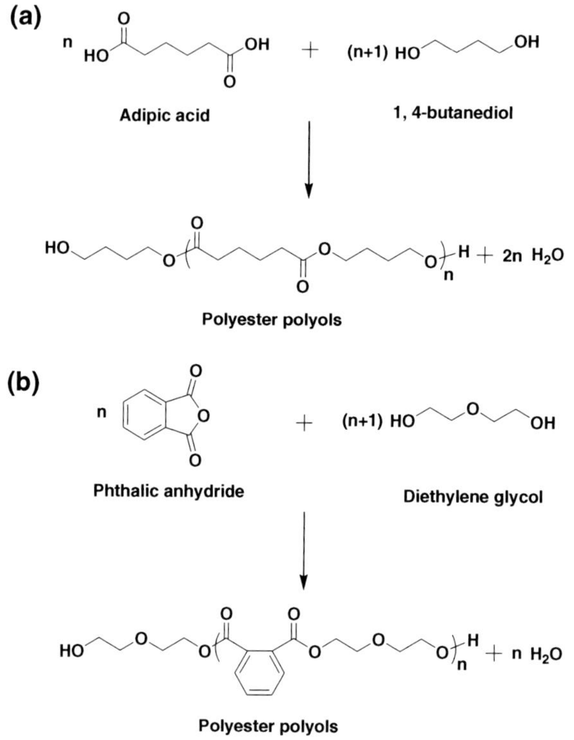

(a)

Adipic acid   1, 4-butanediol

Polyester polyols

(b)

Phthalic anhydride   Diethylene glycol

Polyester polyols

Scheme 1.3 Production of polyester polyols from adipic acid and 1,4-butanediol (**a**), and phthalic anhydride and diethylene glycol (**b**)

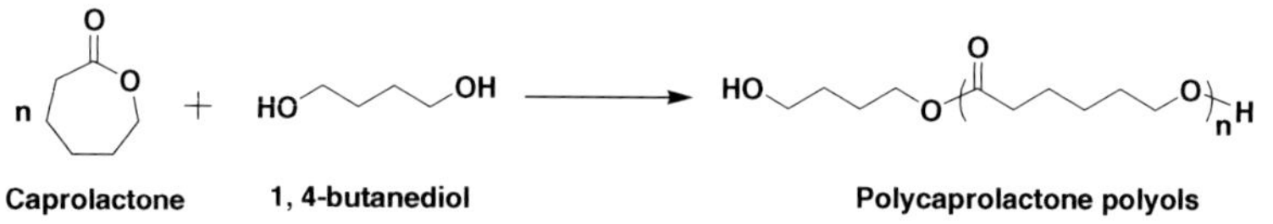

Caprolactone   1, 4-butanediol   Polycaprolactone polyols

Scheme 1.4 Production of polycaprolactone polyols via the initiation of 1,4-butanediol

Recently, concerns about the sustainability of petroleum-based PUs have led to increased demand for bio-based polymeric materials from renewable resources. A large variety of renewable feedstocks, such as vegetable oils [11–14], fatty acids [15], fatty acid methyl esters [16, 17], crude glycerol [18], wood [19], crop residues [19–21], and protein-based feedstocks [22], have been extensively investigated for the production of bio-based polyols. Synthetic methods and properties of bio-based polyols from vegetable oils and their derivatives (Chap. 2), lignocellulosic biomass (Chap. 3), and protein-based feedstocks (Chap. 4), and performance of their derived PUs are discussed in this book.

1.3   Isocyanates

Compared to the diversity of polyols, the choice of isocyanates is relatively limited. Only a few isocyanates are widely used for PU fabrications. Most isocyanates (R–N=C=O) are petroleum-derived and mainly produced from primary amines via phosgenation and subsequent dehydro-halogenation of the resulting carbamoyl chloride (Scheme 1.5).

Isocyanates used for PU production are mainly diisocyanates or polyfunctional isocyanates, which can be aromatic or aliphatic. The major aromatic isocyanates are toluene diisocyanate (TDI), methylene diphenyl diisocyanate (MDI), and polymeric MDI (pMDI) (Fig. 1.1). pMDI is usually a mixture of MDI and MDI oligomers and has a relatively higher functionality (>2) than TDI and MDI [23]. Major aliphatic isocyanates include hexamethylene diisocyanate (HDI), isophorone diisocyanate (IPDI), and dicyclohexyl diisocyanate (HMDI, i.e., hydrogenated MDI) (Fig. 1.2). The reactivity of an isocyanate depends on the positive charge of the carbon atom in the isocyanate group. An electron attracting substituent (R–) attached to an isocyanate group (–NCO) increases the positive charge of the carbon atom in the isocyanate group, and thus improves reactivity of the isocyanate. In contrast, an electron-donating substituent lowers the reactivity of the isocyanate [24]. Therefore, aromatic isocyanates exhibit remarkably higher reactivity than aliphatic ones [8]. The steric hindrance also has an impact on the reactivity of isocyanates. For example, the isocyanate group of 2,4-TDI at the *para*-position is more reactive than that at the *ortho*-position. In industry, aromatic isocyanates are mainly used for the production of PU foams, while aliphatic isocyanates are mainly used for PU coatings.

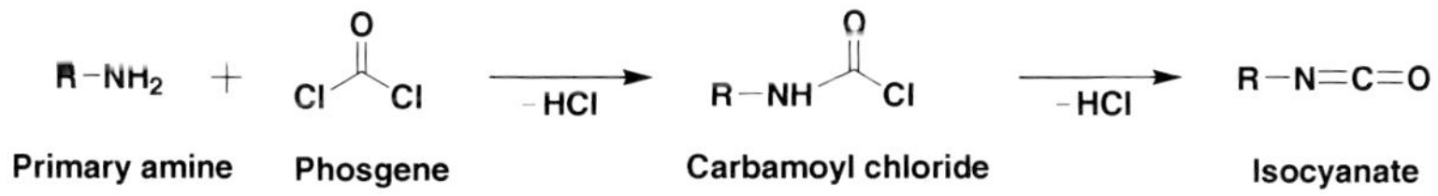

Scheme 1.5   Production of isocyanates from primary amines via phosgenation

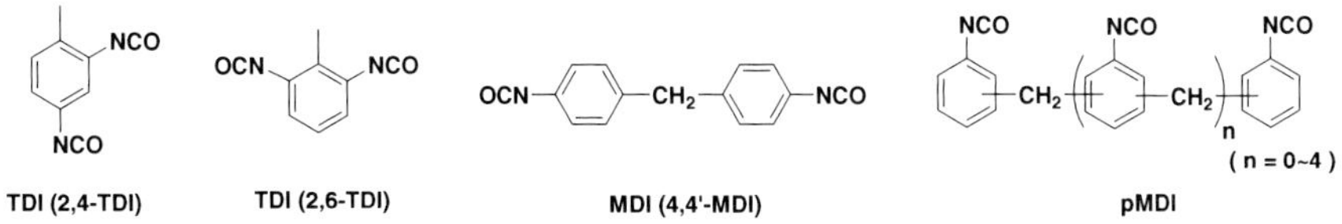

Fig. 1.1   Chemical structures of TDI, MDI, and pMDI

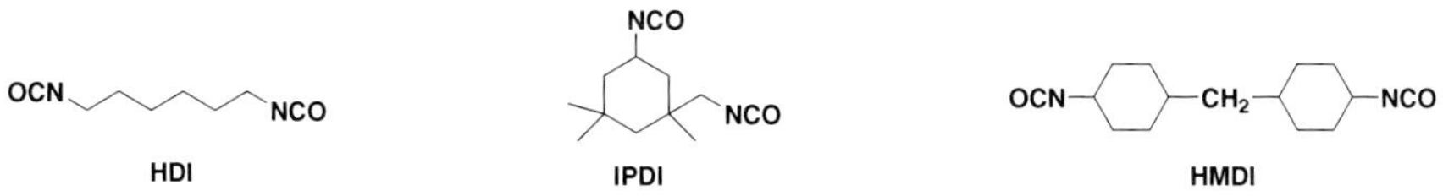

Fig. 1.2   Chemical structures of HDI, IPDI, and HMDI

Currently, nearly all bio-based PUs have been produced from bio-based polyols and petroleum-based isocyanates [25]. Research has indicated that these bio-based PUs show properties comparable to those of petroleum-based PUs [26]. Despite considerable research being focused on the development of polyols from renewable resources, there has been increasing interest from both industry and academia in the synthesis of fatty acid- and vegetable oil-based isocyanates for producing bio-based PUs. Henkel Corporation has commercialized a dimer fatty acid isocyanate with 36 carbon atoms, which has been used for the development of ultraviolet curable optical fiber coatings [27, 28]. General Mills Inc. has synthesized a series of fatty acid-based diisocyanates by multi-step modification of fatty acids to obtain fatty acid-based diamines followed by phosgenation [29].

Academic researchers are pursuing non-phosgene methods for bio-based isocyanate production due to the high toxicity of phosgene. Hojabri et al. [25, 30] synthesized two linear oleic acid-based diisocyanates with and without a carbon-carbon double bond [i.e., 1,16-diisocyanatohexadec-8-ene (HDEDI) and 1,7-heptamethylene diisocyanate (HPMDI)] via Curtius rearrangement (Scheme 1.6). It was demonstrated that oleic acid-based isocyanates had good potential for producing PUs with properties comparable to those produced from petroleum-based isocyanates, such as HDI. For the HDEDI, the presence of a carbon-carbon double bond would provide an extra reaction site for the resulting PUs, which makes vulcanization or cross-linking to enhance the mechanical properties of polymers

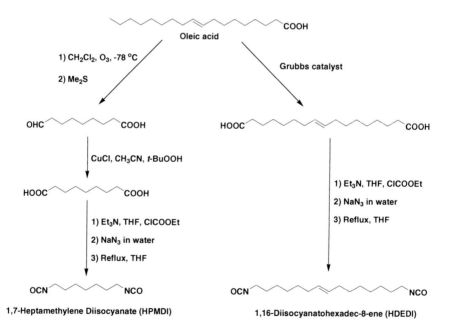

Scheme 1.6 Synthesis of linear saturated and unsaturated terminal diisocyanates from oleic acid [25, 30]

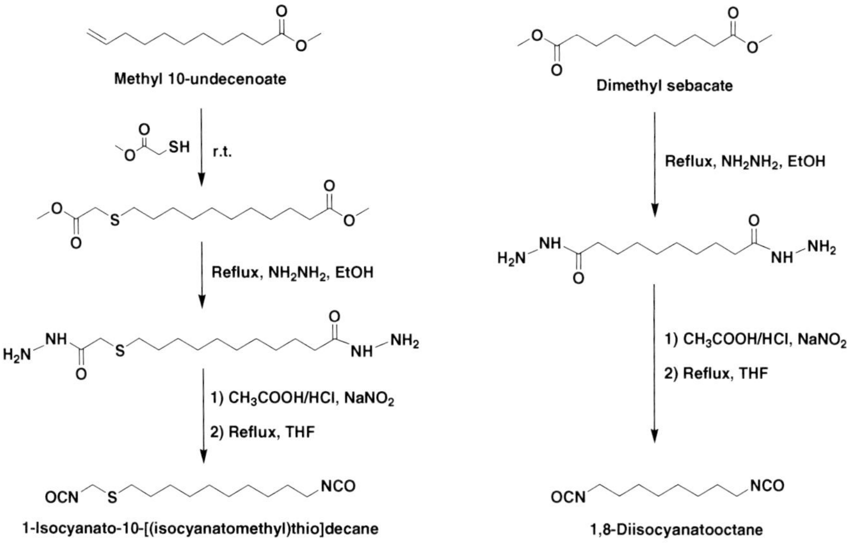

Scheme 1.7 Synthesis of linear diisocyanates from methyl 10-undecenoate and dimethyl sebacate [31]

possible. Two new linear diisocyanates were synthesized using two castor oil derivatives, i.e., methyl 10-undecenoate and dimethyl sebacate, via a series of reactions to convert methyl ester of fatty acids to isocyanate groups [31] (Scheme 1.7). The produced PUs from these two diisocyanates and diols, such as propanediol, butanediol, hexanediol, and isosorbide, were semi-crystalline or amorphous and showed fair thermal stability. Soybean oil-based polyisocyanates

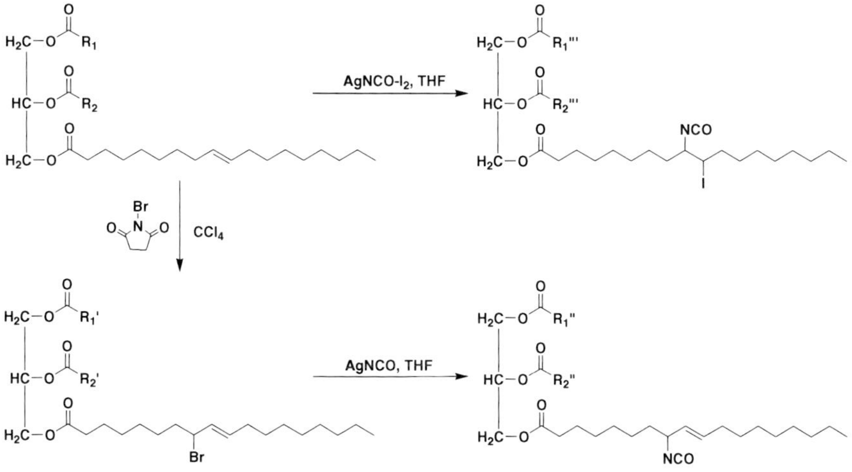

Scheme 1.8 Synthesis of soybean oil-derived polyisocyanates [32, 33]

with functionalities of 2.1 and 3.1 were also prepared by allylic bromination of soybean oil followed by the substitution with AgNCO and by an addition reaction of double bonds in soybean oil with iodine isocyanate generated from AgNCO and iodine, respectfully [32, 33] (Scheme 1.8). Compared to the linear fatty acid-derived terminal diisocyanates, soybean oil-based polyisocyanates have lower reactivity because their isocyanate groups are located in the middle of fatty acid chains. In addition, soybean oil-based polyisocyanates have a plasticizing effect on PU performance due to their multiple dangling alkyl chains. As a result, PUs using soybean oil-based polyisocyanates exhibited lower tensile strength but higher elongation at break than those using linear fatty acid-derived diisocyanates [25].

1.4   Polyurethanes

PUs are polymers that contain multiple urethane bonds (–NHCOO–) and are generally produced by polyaddition reactions between polyols and isocyanates (i.e., diisocyanate or polyisocyanate) (Scheme 1.9). During the PU production process, other reactions may occur, such as isocyanate reactions with water, amine, urethane, urea, and carboxylic acid [7, 8, 10, 34]. Water is commonly used as a blowing reagent in PU foam formulation due to the generation of carbon dioxide during the reaction of isocyanate with water (Scheme 1.10a). Amines, except for tertiary amines, can react with isocyanates to produce ureas (Scheme 1.10b) and generally have higher reactivities than alcohols. Under high temperature (>100 °C), reversible reactions between isocyanates and urea/urethane occur for the formation of biuret/allophanate (Scheme 1.10c, d). Carboxylic acid can also react with isocyanates to generate amides and carbon dioxide (Scheme 1.10e). High acid content is usually avoided in the production of PU foams as it can reduce the catalytic activity of tertiary amines via the formation of ammonium salts [7]. The dimerization and trimerization of isocyanates can also occur in the PU production process [8, 10].

PUs can be linear or cross-linked depending on the functionality of the polyols and isocyanates. Linear PUs are formed when both the polyol and isocyanate contain two functional groups, while cross-linked PUs are formed when either the polyol or isocyanate contains three or more functional groups. In chemical structures of PUs, polyols are commonly called soft segments and contribute to the flexibility of PU chains, whereas isocyanates and chain extenders form hard segments and provide PU chains with rigidity [8]. Therefore, PUs are considered to be copolymers consisting of alternating hard and soft segments [32]. Due to the

Scheme 1.9 Production of PU by polyaddition of polyol and isocyanate

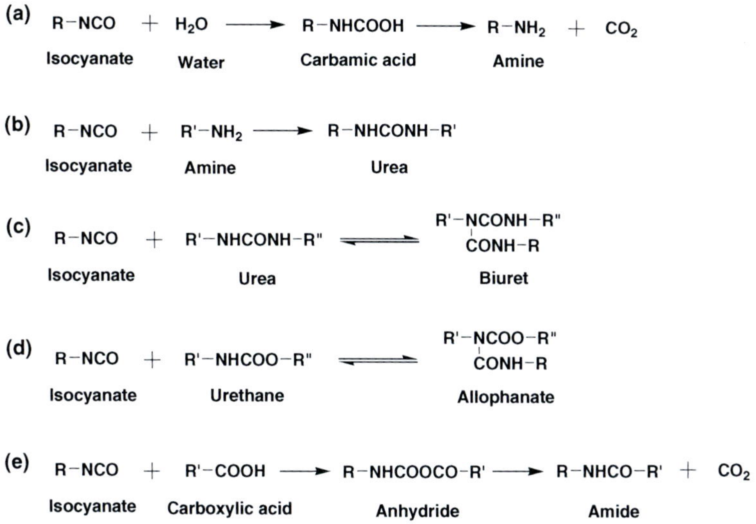

(a) R–NCO + H$_2$O ⟶ R–NHCOOH ⟶ R–NH$_2$ + CO$_2$

Isocyanate | Water | Carbamic acid | Amine

(b) R–NCO + R'–NH$_2$ ⟶ R–NHCONH–R'

Isocyanate | Amine | Urea

(c) R–NCO + R'–NHCONH–R'' ⇌ R'–NCONH–R'' | CONH–R

Isocyanate | Urea | Biuret

(d) R–NCO + R'–NHCOO–R'' ⇌ R'–NCOO–R'' | CONH–R

Isocyanate | Urethane | Allophanate

(e) R–NCO + R'–COOH ⟶ R–NHCOOCO–R' ⟶ R–NHCO–R' + CO$_2$

Isocyanate | Carboxylic acid | Anhydride | Amide

Scheme 1.10 Reactions of isocyanate with water (**a**), amine (**b**), urea (**c**), urethane (**d**), and carboxylic acid (**e**) [7, 8, 10, 34]

availability of a variety of polyols and isocyanates, the properties of PUs can be easily tailored by the types of polyols and isocyanates used in the formula [11]. For example, mechanical and thermal properties of PUs are commonly improved by the incorporation of aromatic segments to the PU backbones via using aromatic isocyanates to replace aliphatic isocyanates [35]. Using the same isocyanates, the tensile strength of PUs from different diols were shown to decrease in the following order: 1,4-butanediol > ethylene diol > 1,5-pentane diol > 1,6-hexamethylene diol > 1,3-propanediol [33]. PUs are commonly characterized according to mechanical and thermal properties. For specific end uses, additional performance, such as thermal conductivity for PU insulation boards, water and organic solvent resistance properties and/or weathering properties for PU coatings, is evaluated.

1.4.1 Foams

PU foams are usually produced by mixing an A-side (i.e., isocyanate) with a B-side in which the polyol, catalyst, blowing agent, surfactant, and other additives are pre-mixed. The most commonly used catalysts are tertiary amines and organometallics, primarily tin-containing compounds. The blowing agent is generally water or a hydrocarbon such as pentane. Silicone-containing compounds such as polydimethylsiloxane and polyoxyalkylene-polysiloxane copolymers are generally used as surfactants. Other additives such as chain extenders, crosslinkers, fire retardants, and fillers are also used in PU formulations to obtain foams with improved performance and/or special properties. PU foams can either be rigid or flexible. Their typical production formulas are shown in Table 1.1. The cellular structure and morphology of both rigid and flexible foams are characterized with microscopy such as scanning electron microscopy (SEM), scattering such as small angle X-ray scattering (SAXS), or thermal analysis such as dynamic mechanical analysis (DMA) and differential scanning calorimetry (DSC) [10]. The thermal stability of foams can be characterized by thermogravimetric analysis (TGA). For the performance of rigid foams, properties such as density, compressive strength, thermal conductivity, closed cell content, and dimensional stability are generally evaluated; while properties such as density, tensile strength, resilience, compression set, and indentation force-deflection are tested for flexible foams (Table 1.2).

Table 1.1 Typical production formulas of rigid and flexible PU foams [36]

Rigid PU foams		Flexible PU foams	
Ingredients	Parts by weight	Ingredients	Parts by weight
Polyol	100	Polyol	100
Polycat 5	1.26	Stannous octoate	0.3
Polycat 8	0.84	Dibutyltin dilaurate	0.3
Dabco DC5357	2.5	Dabco 33-LV	0.6
Distilled water	3.0	Dabco BL-17	0.2
		Dietanolamine	2.2
		Dabco DC2585	1.0
		Distilled water	5.0
pMDI (PAPI 27)	Index 110[a]	pMDI (PAPI 27)	Index 80[a]

Note [a]Isocyanate index, which is the ratio of actual weight to theoretical weight of isocyanate, multiplied by 100

Table 1.2 Typical properties of petroleum-based rigid and flexible PU foams [8]

Rigid PU foams			Flexible PU foams		
Density (kg/m^3)		24–32	Density (kg/m^3)		17.6–36.8
Compressive strength (kPa)		137–310	Tensile strength (kPa)		62.1–227.5
Thermal conductivity (mW m^{-1} K^{-1})		24.0–28.8	Resilience (%)		15–70
Closed cell content (%)		92–98	Compression set (%) at 50 % strain		1–7
Dimensional stability (%)	70 °C, 100 % relative humidity	7–15	Indentation force deflection (kPa)	At 25 % compression	1.2–6.2
	100 °C, ambient	5–10		At 65 % compression	2.2–11.7

Table 1.3 Performances of PU coatings from polyether and polyester polyols [10]

	PU coatings from polyether polyols	PU coatings from polyester polyols
Viscosity	Low	High
Appearance	Good	Very good
Hardness	Soft	Medium
Brittleness	Excellent	Excellent
Gloss	Poor	Fair
Organic solvent resistance	Fair	Fair

1.4.2   Coatings

PU coatings have been widely used in various applications such as automotive, electronics, wood products, and machinery because of their excellent chemical, solvent, and abrasion resistance, as well as toughness combined with good low-temperature flexibility [37]. Technologies developed for PU coatings include two-component, non-isocyanate reactive, and one-component systems, with the first two being predominately used [8, 10]. Two-component systems are composed of a polyol side and an isocyanate side, which are mixed prior to application and then cured at ambient conditions. In two-component PU coatings, oven-cured coatings can be produced by mixing a polyol with a blocked isocyanate before curing. As it is heated, the blocked isocyanate is de-blocked and the isocyanate groups are exposed to react with the polyol. Non-isocyanate reactive systems do not require further reaction with isocyanates during applications, such as waterborne PU dispersions. One-component systems consist of prepolymers with low isocyanate contents, which are cured by a reaction with moisture. Thus, they are often called moisture-cured systems [8]. Similar to PU foams, the morphology and thermal properties of PU coatings are commonly characterized by SEM, DMA and/or DSC, and TGA. Properties, including adhesion, bending, hardness, water and organic solvent resistance, viscosity, and tensile strength, are generally evaluated for coating applications. Depending on the polyol structure, PU coatings from polyether and polyester polyols exhibit various performances (Table 1.3) [10].

References

1. Bayer O, Siefken W, Rinke H, Orthner L, Schild H (1937) A process for the production of polyurethanes and polyureas. German Patent DRP 728981
2. Marketsandmarkets (2013) Polyurethane (PU) foams market by types (rigid & flexible), end-user industries (bedding & furniture, building & construction, electronics, automotives, footwear, packaging, & others), & geography (North America, West Europe, Asia-Pacific & Row)-global trends & forecasts to 2018. http://www.marketsandmarkets.com/PressReleases/polyurethane-foams.asp. Accessed Mar 2015

3. Mahendran AR, Aust N, Wuzella G, Müller U, Kandelbauer A (2012) Bio-based non-isocyanate urethane derived from plant oil. J Polym Environ 20:926–931
4. Fleischer M, Blattmann H, Mülhaupt R (2013) Glycerol-, pentaerythritol-and trimethylolpropane-based polyurethanes and their cellulose carbonate composites prepared via the non-isocyanate route with catalytic carbon dioxide fixation. Green Chem 15:934–942
5. Palaskar DV, Boyer A, Cloutet E, Alfos C, Cramail H (2010) Synthesis of biobased polyurethane from oleic and ricinoleic acids as the renewable resources via the AB-type self-condensation approach. Biomacromolecules 11:1202–1211
6. Deepa P, Jayakannan M (2008) Solvent-free and nonisocyanate melt transurethane reaction for aliphatic polyurethanes and mechanistic aspects. J polym Sci A Polym Chem 46:2445–2458
7. Ionescu M (2005) Chemistry and technology of polyols for polyurethanes. Rapra Technology, Ltd., Shropshire
8. Szycher M (1999) Szycher's handbook of polyurethanes. CRC Press, Florida
9. Fink JK (2013) Reactive polymers fundamentals and applications: a concise guide to industrial polymers, 2nd edn. Elsevier, Amsterdam
10. Randall D, Lee S (2002) The polyurethanes book. John Wiley & Sons Ltd., UK
11. Pfister DP, Xia Y, Larock RC (2011) Recent advances in vegetable oil-based polyurethanes. ChemSusChem 4:703–717
12. Petrović ZS (2008) Polyurethanes from vegetable oils. Polym Rev 48:109–155
13. Babb DA (2012) Polyurethanes from renewable resources. Synthetic biodegradable polymers. Springer, Berlin, pp 315–360
14. Lligadas G, Ronda JC, Galià M, Cádiz V (2010) Plant oils as platform chemicals for polyurethane synthesis: current state-of-the-art. Biomacromolecules 11:2825–2835
15. Lligadas G, Ronda JC, Galiá M, Cádiz V (2010) Oleic and undecylenic acids as renewable feedstocks in the synthesis of polyols and polyurethanes. Polymer 2:440–453
16. Lligadas G, Ronda JC, Galià M, Biermann U, Metzger JO (2006) Synthesis and characterization of polyurethanes from epoxidized methyl oleate based polyether polyols as renewable resources. J polym Sci A Polym Chem 44:634–645
17. Petrović ZS, Cvetković I, Milić J, Hong D, Javni I (2012) Hyperbranched polyols from hydroformylated methyl soyate. J Appl Polym Sci 125:2920–2928
18. Luo X, Hu S, Zhang X, Li Y (2013) Thermochemical conversion of crude glycerol to biopolyols for the production of polyurethane foams. Bioresour Technol 139:323–329
19. Hu S, Luo X, Li Y (2014) Polyols and polyurethanes from the liquefaction of lignocellulosic biomass. ChemSusChem 7:66–72
20. Aniceto JP, Portugal I, Silva CM (2012) Biomass-based polyols through oxypropylation reaction. ChemSusChem 5:1358–1368
21. Hu S, Wan C, Li Y (2012) Production and characterization of biopolyols and polyurethane foams from crude glycerol based liquefaction of soybean straw. Bioresour Technol 103:227–233
22. Mu Y, Wan X, Han Z, Peng Y, Zhong S (2012) Rigid polyurethane foams based on activated soybean meal. J Appl Polym Sci 124:4331–4338
23. DFG (2012) 4,4-Methylene diphenyl isocyanate (MDI) and polymeric MDI (PMDI) [MAK Value Documentation, 1997]. The MAK-Collection for Occupational Health and Safety Wiley-VCH Verlag GmbH & Co. KGaA
24. Sharmin E, Zafar F (2012) Polyurethane: an introduction. InTech. Open Access Publisher
25. Hojabri L, Kong X, Narine SS (2010) Novel long chain unsaturated diisocyanate from fatty acid: synthesis, characterization, and application in bio-based polyurethane. J polym Sci A Polym Chem 48:3302–3310
26. Javni I, Zhang W, Petrović ZS (2003) Effect of different isocyanates on the properties of soy-based polyurethanes. J Appl Polym Sci 88:2912–2916
27. Coady CJ, Krajewski JJ, Bishop TE (1986) Polyacrylated oligomers in ultraviolet curable optical fiber coatings. US Patent 4,608,409
28. Bishop TE, Coady CJ, Zimmerman JM (1986) Ultraviolet curable buffer coatings for optical glass fiber based on long chain oxyalkylene diamines. US Patent 4,609,718

29. Kamal MR, Kuder RC (1972) Diisocyanates. US Patent 3,691,225
30. Hojabri L, Kong X, Narine SS (2009) Fatty acid-derived diisocyanate and biobased polyurethane produced from vegetable oil: synthesis, polymerization, and characterization. Biomacromolecules 10:884–891
31. More AS, Lebarbé T, Maisonneuve L, Gadenne B, Alfos C, Cramail H (2013) Novel fatty acid based di-isocyanates towards the synthesis of thermoplastic polyurethanes. Eur Polym J 49:823–833
32. Çayli G, Küsefoğlu S (2008) Biobased polyisocyanates from plant oil triglycerides: synthesis, polymerization, and characterization. J Appl Polym Sci 109:2948–2955
33. Çayli G, Küsefoğlu S (2010) A simple one-step synthesis and polymerization of plant oil triglyceride iodo isocyanates. J Appl Polym Sci 116:2433–2440
34. Delebecq E, Pascault J-P, Boutevin B, Ganachaud F (2012) On the versatility of urethane/urea bonds: reversibility, blocked isocyanate, and non-isocyanate polyurethane. Chem Rev 113: 80–118
35. Pandya MV, Deshpande DD, Hundiwale DG (1986) Effect of diisocyanate structure on viscoelastic, thermal, mechanical and electrical properties of cast polyurethanes. J Appl Polym Sci 32:4959–4969
36. Tu Y-C, Suppes GJ, Hsieh F-H (2008) Water-blown rigid and flexible polyurethane foams containing epoxidized soybean oil triglycerides. J Appl Polym Sci 109:537–544
37. Coutinho F, Delpech MC, Alves LS (2001) Anionic waterborne polyurethane dispersions based on hydroxyl-terminated polybutadiene and poly (propylene glycol): synthesis and characterization. J Appl Polym Sci 80:566–572

Chapter 2
Polyols and Polyurethanes from Vegetable Oils and Their Derivatives

Abstract Vegetable oils and their derivatives have been widely used for the production of various polymers including polyols and polyurethanes. Vegetable oil derivatives, such as fatty acids, fatty acid esters, and crude glycerol, can be obtained via hydrolysis or transesterification of vegetable oils. Polyols and polyurethanes with properties comparable to those of petroleum-based analogs have been prepared from vegetable oils and their derivatives for various applications such as foams, coatings, adhesives, etc. This chapter reviews the structures and compositions of vegetable oils and their derivatives, synthetic methods of producing polyols from vegetable oils and their derivatives, properties of these polyols, and performance and applications of the resulting polyurethanes.

Keywords Bio-based polyols · Polyurethanes · Vegetable oils · Fatty acids · Fatty acid esters · Crude glycerol

2.1 Introduction

Vegetable oils are triglycerides, also known as triacylglycerols, which are triesters of glycerol and different fatty acids (Fig. 2.1). Depending on the origin and type of fatty acids in vegetable oils, the fatty acid side chains contain carbon numbers ranging from 8 to 24 and carbon-carbon double bond numbers from 0 to 5 [1], leading to high variability of vegetable oil compositions. As shown in Fig. 2.2, most vegetable oils consist of five major fatty acids: palmitic (C16:0), stearic (C18:0), oleic (C18:1), linoleic (C18:2), and linolenic acid (C18:3), in which for the (Cm:n) designation, Cm indicates the number of carbon atoms and n indicates the number of double bonds [1, 2]. According to the USDA-FAS (United State Department of Agriculture, Foreign Agricultural Service) [3], the annual world production of vegetable oils has been steadily increasing from around 148.8 MMT (million metric tons) in 2010/2011 to 170.9 MMT in 2013/2014. Palm, soybean, rapeseed, and sunflower seed are four predominant types of vegetable oil feedstocks, accounting for approximately 86 % of the global production of vegetable oils [3]. Figure 2.3

© The Author(s) 2015   15
Y. Li et al., *Bio-based Polyols and Polyurethanes*,
SpringerBriefs in Green Chemistry for Sustainability,
DOI 10.1007/978-3-319-21539-6_2

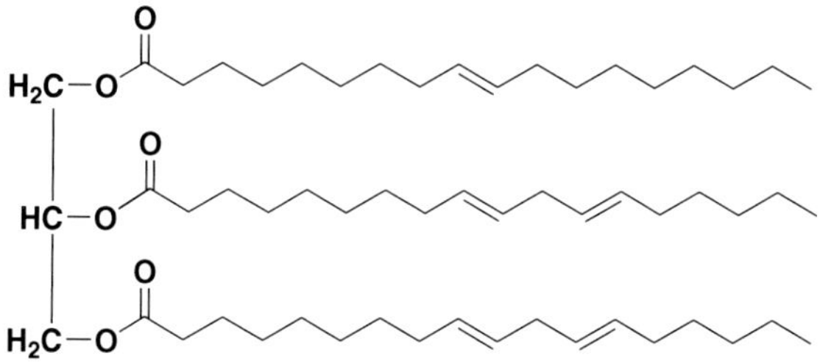

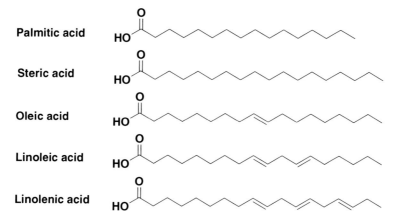

Fig. 2.1 Schematic representation of triglyceride structure of vegetable oils

Palmitic acid

Steric acid

Oleic acid

Linoleic acid

Linolenic acid

Fig. 2.2 Chemical structures of five major fatty acids found in vegetable oils

illustrates the 2013/2014 global annual production distributions of these major vegetable oils, and Table 2.1 shows the typical fatty acid profiles of vegetable oils.

Fatty acids, fatty acid esters, and glycerol are three different derivatives from vegetable oils. Fatty acids and fatty acid esters are usually produced by hydrolysis and transesterification of vegetable oils with water and alcohol, respectively. Scheme 2.1 shows schematic routes for the production of fatty acids and fatty acid esters from vegetable oils. Fatty acid methyl esters (i.e., biodiesel) are one type of important fatty acid ester and are obtained commercially by the transesterification of vegetable oils with methanol under the catalysis of sodium hydroxide or potassium hydroxide. Crude glycerol is a byproduct of the biodiesel production process. It is estimated that approximately 1 kg of crude glycerol is generated for every 10 kg biodiesel produced [5]. Compared to glycerol, crude glycerol has a significantly different composition and contains multiple impurities such as methanol, water, fatty acids, fatty acid methyl esters, soap, and glycerides [6]. The rapid growth in the production of biodiesel worldwide has generated large volumes of crude glycerol. Due to the high cost of refining it, especially for small- or medium-sized biodiesel producers [7], there has been extensive interest in the development of feasible biological or chemical approaches for converting low-value

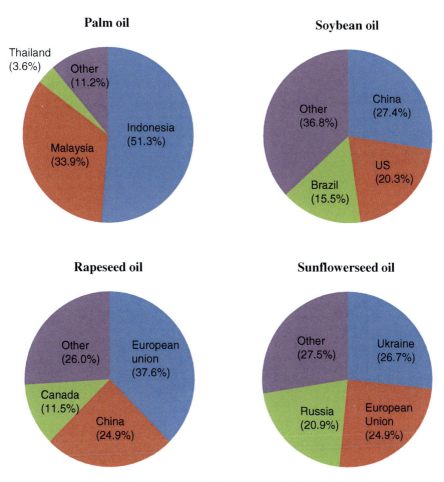

Fig. 2.3 Global annual production distributions (2013/2014) of palm oil, soybean oil, rapeseed oil, and sunflowerseed oil [3]

Table 2.1 Typical fatty acid profiles of major vegetable oils [1, 4]

Vegetable oils	Fatty acid profiles (wt%)				
	C16:0	C18:0	C18:1	C18:2	C18:3
Palm oil	44.0	4.5	39.2	10.1	0.4
Soybean oil	11.3	3.4	23.1	55.8	6.4
Rapeseed oil	4.0	2.0	56.0	26.0	10.0
Sunflowerseed oil	5.9 (3.8)[a]	4.4 (4.1)	19 (78.4)	67.5 (11.3)	2.9 (trace)

[a]Values in parenthesis are for high-oleic sunflowerseed oil

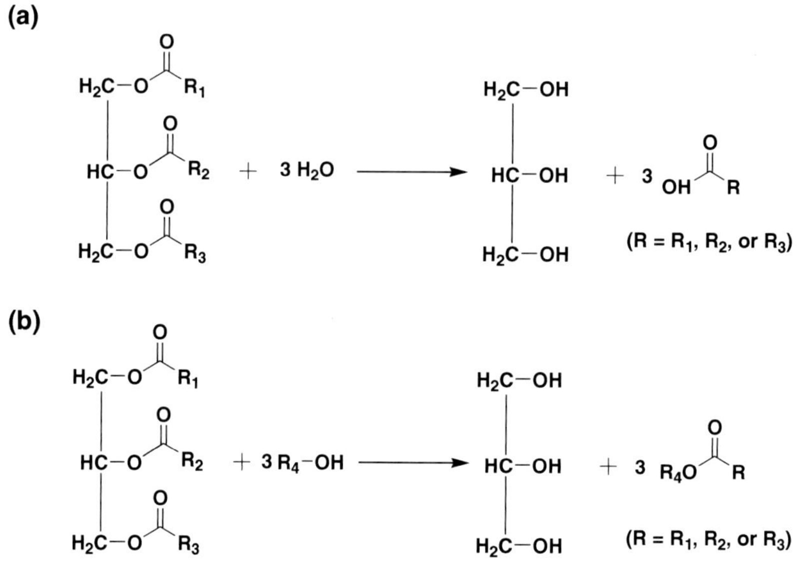

(a)

(b)

Scheme 2.1 Schematic routes for the production of fatty acids (**a**) and fatty acid esters (**b**) from vegetable oils (R_1, R_2, and R_3 are fatty acid chains and can be different or identical; R_4 is an alkyl group of an alcohol)

crude glycerol (approximately $0.1/kg [5]) to value-added products, including polyols and polyurethanes (PUs) [8–12].

Natural vegetable oils, with the exception of castor and lesquerella oils, do not contain hydroxyl groups. When used as feedstocks for polyol production, vegetable oils are often chemically modified to introduce hydroxyl groups into their structures. Carbon-carbon double bonds and ester linkages are two major functional moieties present in structures of vegetable oils. Almost all synthetic routes for vegetable oil-based polyol production, including epoxidation followed by oxirane ring-opening, hydroformylation followed by hydrogenation, ozonolysis, thiol-ene coupling, transesterification, and amidation, start from one of these two functional moieties. Fatty acid esters and fatty acids have functional moieties similar to vegetable oils, such as carbon-carbon double bonds and ester linkages/carboxyl groups. The methods used for producing vegetable oil-based polyols can also be used to produce polyols from fatty acid esters or fatty acids. Besides these methods, other methods, such as dimerization of fatty acids and hydrosilylation of fatty acid esters, have been developed to functionalize fatty acids or fatty acid esters for polyol synthesis. The presence of multiple reactive components in crude glycerol allows for esterification and transesterification by which polyols can be produced for PU applications. This chapter mainly focuses on the methods commonly used for the preparation of polyols from vegetable oils and their derivatives such as fatty acids, fatty acid esters, and crude glycerol. The polyol properties and their derived PUs are also discussed.

2.2 Vegetable Oil-Based Polyols and Polyurethanes

There are five different methods typically used to produce polyols from vegetable oils. These processes and the characteristics of the derived polyols and PUs are discussed in the following sections.

2.2.1 Epoxidation and Oxirane Ring-Opening Pathway

Epoxidation has been one of the most commonly used methods for the functionalization of carbon-carbon double bonds [13]. The epoxidation of vegetable oils can be conducted either in bulk or in solution with preformed or in situ prepared peracids, i.e. an active oxygen provider, under either homogenous or heterogeneous catalysis [14]. The epoxidation is usually conducted at a temperature between 30 and 80 °C for a reaction time of 10–20 h, depending on the types of feedstocks and ratios of reactants involved in the reaction. Under optimized conditions, conversion yields higher than 90 % can be achieved [15–17]. The undesirable side reactions of oxirane ring-opening during epoxidation can be largely minimized by conducting the reaction in a solution and at low temperature as well as under acidic ion exchange resin or lipase catalysis [14, 15, 18–21]. Polyols are produced from epoxidized vegetable oil by oxirane ring-opening reactions using a broad range of active hydrogen containing compounds such as alcohols, inorganic and organic acids, amines, water, and hydrogen [13, 19, 22–26]. A schematic representation of the production of polyols from vegetable oils by epoxidation followed by oxirane ring-opening is shown in Scheme 2.2. Usually, the epoxidation and ring-opening reactions are conducted in two separate steps, although a one-step process combining epoxidiation and ring-opening reactions has been reported [27–29].

The properties of polyols produced by epoxidation and subsequent oxirane ring-opening depend on several production variables including feedstock characteristics and the types of ring-opening agents. Vegetable oils with a higher degree of unsaturation produce polyols with higher hydroxyl functionalities, resulting in PUs with higher crosslinking density and higher tensile strength [30]. Oxirane ring-opening agents are divided into three major categories: (a) **Alcohols**. When monoalcohols are used for ring-opening, each epoxy moiety only generates one secondary hydroxyl group, which are much less reactive than primary hydroxyl groups [31]. Methanol is the common choice of monoalcohols for ring-opening due to its low cost, low molecular weight, and low boiling point [1]. In order to produce polyols with higher functionalities and with primary hydroxyl groups, diols such as 1,2-propanediol and ethylene glycol have also been used for oxirane ring-opening reactions [22, 25]. However, the produced polyols tend to possess high viscosities due to their increased hydroxyl numbers. (b) **Acids**. Carboxylic acids such as formic and acetic acids have been used for ring-opening of epoxidized vegetable oils to produce polyester polyols, which have been shown to have potential for

Scheme 2.2 Production of vegetable oil-based polyols via epoxidation and oxirane ring-opening pathway (R$_1$ and R$_2$ are fatty acid side chains of vegetable oil; R$_1$′, R$_2$′, R$_1$″, and R$_2$″ are modified fatty acid side chains of vegetable oil)

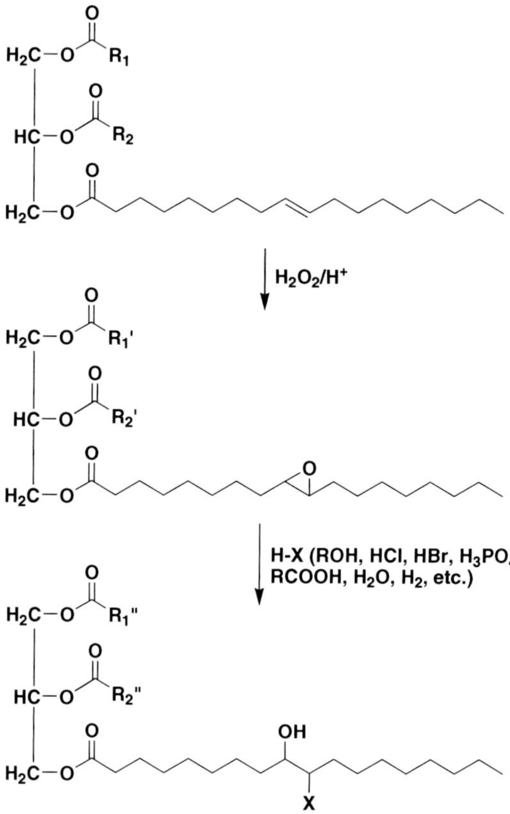

anti-wear applications [13, 26, 27]. Inorganic acids such as HCl, HBr, and H$_3$PO$_4$ have also been reported as ring-opening agents [23, 32]. Due to the incompatibility between inorganic acids and epoxidized vegetable oils, polar organic solvents such as acetone and *t*-butyl alcohol are typically added into the reactor to facilitate reactions. There are some drawbacks associated with the polyols from oxirane ring-opening reactions of epoxidized vegetable oils by inorganic acids. Polyols produced from oxirane ring-opening by HCl and HBr are waxes at room temperature [23], while those produced by H$_3$PO$_4$ contain significant fractions of oligomers because of the oligomerizations that occur between oxirane groups [32]. (c) **Hydrogen**. Polyols from ring-opening of epoxidized vegetable oils by hydrogen under Raney nickel catalysis are grease at room temperature, which has limited their applications in preparing PUs [23]. Table 2.2 lists properties of polyols produced from epoxidized soybean oil using different oxirane ring-opening agents.

Simultaneous oxirane ring-opening and transesterification [16] of epoxidized canola oil was achieved by using a strong acid catalyst (e.g., sulfuric acid) and excess amounts of diols (e.g., 1,2-propanediol or 1,3-propanediol). Because transesterification effectively removed the glycerol backbone from polyols, the

Table 2.2 Properties of polyols produced from epoxidized soybean oil using different oxirane ring-opening agents

Ring-opening agents	Polyol properties						References
	Hydroxyl number (mg KOH/g)	Hydroxyl group type	Acid number (mg KOH/g)	f_n^c	Viscosity (Pa.s)	Molecular weight (g/mol)	
Methanol	199	Secondary	$-^b$	3.7	12 $(25\ ^\circ C)^d$	1053	[23]
	180	Secondary	–	–	0.6 (45 °C)	–	[22]
	148–174	Secondary	–	2.6–3.2	–	1001–1025[a]	[25]
1,2-Ethanediol	253	Primary, secondary	–	–	1 (45 °C)	–	[22]
	187–226	Primary, secondary	–	3.4–4.2	–	1005–1038[a]	[25]
1,2-Propanediol	289	Primary, secondary	–	–	1 (45 °C)	–	[22]
	211–237	Primary, secondary	–	3.8–4.6	–	1010–1084[a]	[25]
Lactic acid	210[a]	Secondary	–	4.2	–	1120	[19]
	171[a]	Secondary	3.6	5.3	47[e]	1738[a]	[26]
Glycolic acid	203[a]	Primary, secondary	2.6	4.9	221[e]	1352[a]	[26]
Acetic acid	188[a]	Secondary	1.8	4.3	55[e]	1281[a]	[26]
Formic acid	104–162	Secondary	1.8–2.5	1.9–3.2	3–10 (30 °C)	1027–1086	[27]
Linoleic acid	76–112	Secondary	4–25	–	1.4–2.8 (22 °C)	–	[24]
Ricinoleic acid	152–163	Secondary	5–16	–	7.7–9.4 (22 °C)	–	[24]
Hydrochloric acid	197	Secondary	–	3.8	Grease[f]	1071	[23]
Hydrobromic acid	182	Secondary	–	4.1	Grease[f]	1274	[23]
Phosphoric acid	153–253	Secondary	1.4–48	12.8–17.5[a]	3.2–5.3	3870–4700	[32]
Hydrogen	212–225	Secondary	–	3.5–3.8	Grease[f]	938–947	[23, 34]

[a]Calculated values
[b]Not reported
[c]Functionality
[d]Testing temperature
[e]Testing temperature not reported
[f]At room temperature

produced polyols had lower molecular weights (ca. 433 g/mol) and viscosities (about 3 Pa.s), and higher hydroxyl numbers (about 270 and 320 mg KOH/g) than polyols obtained by oxirane ring-opening alone [16]. In a more recent study, polyols with similar properties have also been produced from epoxidized soybean oil through simultaneous ring-opening and amidation reactions [33].

Epoxidized polyols (polyols derived from epoxidation and subsequent oxirane ring-opening) have been used for the preparation of many PU products such as resins, foams, and coatings. By controlling factors such as vegetable oil composition, ring-opening agent, and degree of epoxidation, polyols and PUs with varying properties have been produced. A close relationship between the properties of epoxidized polyols from vegetable oils and their derived PUs has been observed: the higher the hydroxyl number and functionality of the polyol, the higher the crosslinking density, the T_g, and the tensile strength of the PUs [25, 30, 35]. Halogenated (e.g., chlorinated and brominated) PUs possessed higher mechanical properties, higher T_g, and lower linear thermal expansion coefficients than non-halogenated ones (e.g., methoxylated and hydrogenated PUs), which could be attributed to their stronger intermolecular attractions, higher crosslinking densities, and lower free volumes resulted from large halogen atoms; however, compared to non-halogenated PUs, halogenated ones showed lower thermal stability and higher initial weight loss as a result of the dissociation of bromine or chlorine [36]. The effects of the structural heterogeneity of vegetable oils on the properties of epoxidized polyols and PUs were evaluated by preparing polyols from soybean oil with different hydroxyl numbers via partial hydrogenation of epoxy rings [34]. The negative effects of structural heterogeneity on PU properties were found to be more pronounced for PUs with lower crosslinking densities produced from polyols with lower hydroxyl numbers. In addition, the properties of PUs can be manipulated by the NCO/OH ratio. In one typical study, PU networks ranging from elastomeric to glassy plastics were obtained for NCO/OH ratios ranging from 0.4 to 1.05 [37].

Rigid and flexible PU foams have been prepared from epoxidized polyols derived from vegetable oils such as soybean oil and rapeseed oil. Compared to petroleum-derived polyols that contain mostly primary hydroxyl groups, epoxidized polyols from vegetable oils have lower reactivity and need longer curing time when they react with isocyanates for PU foam production.

Several strategies, such as addition of a crosslinker [38], post-modifications by alcoholysis [29], and mixing with commercial petroleum-derived polyols [39, 40], have been developed to produce rigid PU foams from epoxidized polyols. In view of its high hydroxyl number (1829 mg KOH/g) and compact backbone, glycerol has been proven to be an excellent crosslinker to increase the rigidity of epoxidized polyol-based PU foams [38]. Under preferred glycerol addition (10–25 wt% of soy polyol) and optimized foam formulation conditions, the produced rigid PU foams exhibited mechanical and thermal-insulating properties comparable to analogues prepared from commercial polyether polyols. Applying triethanolamine-based alcoholysis to post-modify hydroxylated rapeseed oil effectively increased polyol hydroxyl numbers from 100 to 367 mg KOH/g, and the resulting PU foams showed very similar mechanical and thermal-insulating properties to those prepared from a commercial petroleum-derived polyol [29]. Partial substitution of commercial petroleum-derived polyols by epoxidized polyols has been shown to be feasible for the production of high-quality rigid PU foams [39, 40]. Under optimal foaming conditions, PU foams produced with up to 50 % epoxidized polyol substitution possessed mechanical and thermal-insulating properties comparable with those based on 100 % petroleum-derived polyols [39, 40]. However, the epoxidized polyol-derived PU foams had the drawback of higher aging rates, i.e. an increase of thermal conductivity with time due to its higher N_2 permeation, than petroleum-derived foams, which may be alleviated by the addition of a crosslinker, such as glycerol [39]. Flexible PU foams from epoxidized polyols have also been prepared by this blending method. Mixing epoxidized polyols (up to 50 % by weight) with petroleum-derived polyols was found to increase the compressive strength and modulus of the resulting foams [41–45]. Several factors can contribute to this phenomenon including the existence of a high glass transition phase rich in vegetable oil-derived polyols, high hard segment concentration, and improved hard domain ordering in foam morphologies [43]. Flexible PU foams from epoxidized polyols also exhibited lower resilience values and higher hysteresis loss due to their decreased elasticity [45]. However, these issues may be alleviated through optimization of foaming formulations, which can be achieved by approaches such as controlling the amount and type of epoxidized polyols incorporated and the NCO/OH ratio [42, 44].

Waterborne PU dispersions, including anionic and cationic PU or hybrid dispersions, have been synthesized from methoxylated soybean oil polyols [46–51]. As the polyol functionality and/or hard segment increased, PU films exhibited increased crosslinking density, T_g, and tensile strength. By employing polyols with increasing functionalities ranging from 2.4 to 4.0, PU films ranged from elastomers to ductile plastics to rigid plastics [46, 49]. Urethane-acrylic and urethane-styrene-acrylic hybrid latexes have been prepared from soybean oil-based epoxidized polyols by emulsion polymerization [47, 50, 51]. Generally, these hybrid latex films showed improved tensile strength, Young's modulus, and thermal stability, compared to non-hybrid PU films [47, 50, 51].

2.2.2 Hydroformylation and Hydrogenation Pathway

Hydroformylation followed by hydrogenation is another important pathway for the preparation of vegetable oil-derived polyols. During preparation, double bonds in vegetable oil structures are first converted to aldehydes via catalyzed hydroformylation by syngas (typically a 1:1 mixture of CO and H_2), and then to hydroxyl groups via the hydrogenation of aldehyde (Scheme 2.3) [52, 53]. Rhodium- or cobalt- based catalysts are the common ones used for the hydroformylation of vegetable oils. Excellent double bond to aldehyde conversion yield (95 %) has been obtained using rhodium catalysts as compared to that of cobalt catalysts (67 %) [52]. However, rhodium catalysts are expensive and, when used as hydroformylation catalysts, the hydrogenation process

Scheme 2.3 Production of vegetable oil-based polyols via hydroformylation and hydrogenation pathway (R_1 and R_2 are fatty acid side chains of vegetable oil; R_1', R_2', R_1'', and R_2'' are modified fatty acid side chains of vegetable oil)

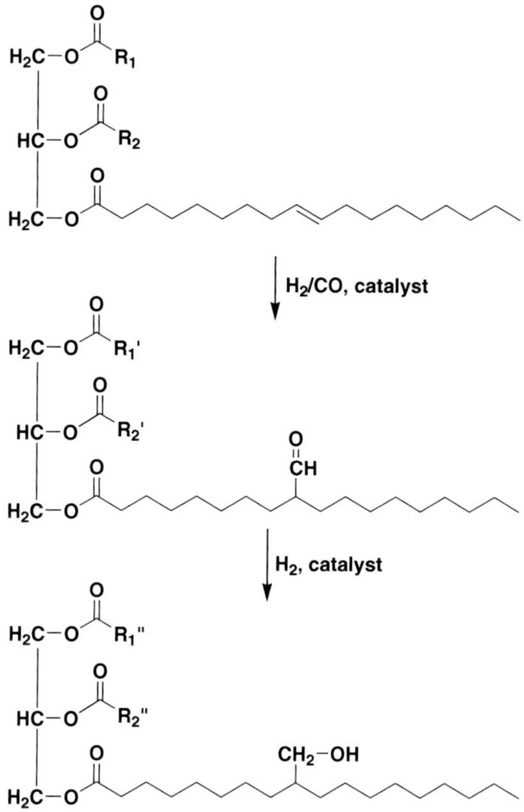

that follows requires an additional Raney nickel catalyst. In contrast, cobalt catalysts are less expensive and capable of catalyzing both hydroformylation and hydrogenation reactions, but they require harsher reaction conditions [1, 52]. A comparison between rhodium- and cobalt- based catalysts revealed that polyols obtained from the rhodium catalyzed process exhibited higher hydroxyl numbers and higher functionalities than those from the cobalt catalyzed process [52]. As a result, rigid plastic-like PUs were produced from rhodium-derived polyols, while hard rubber-like PUs from cobalt-derived polyols [52]. The major advantage of the hydroformylation-hydrogenation process is the formation of primary hydroxyl groups, which are preferred to the secondary hydroxyl groups usually obtained from the epoxidation and oxirane ring-opening pathway. Consequently, polyols produced by the hydroformylation and hydrogenation pathway are more reactive than epoxidized polyols and a smaller amount of catalyst is required for their reactions with isocyanates [68].

As only one hydroxyl group per carbon-carbon double bond is generated via hydroformylation and hydrogenation, the polyols have the same functionalities as the original vegetable oils. This makes the properties of polyols dependent on the compositions of the starting vegetable oils. In order to widen the structural and property versatility of polyols, other structural modifications have been introduced in combination with hydroformylation and hydrogenation, such as methanolysis and polycondensation by which hyperbranched polyols with high molecular weights and functionalities [54] and triols with high molecular weights [55] were prepared.

When reacted with isocyanates for PUs, hydroformylated polyols (polyols derived via the hydroformylation and hydrogenation pathway) showed shorter gel time and better curing efficiency, compared to epoxidized polyols [56]. PU foams with enhanced rigidity can be produced from hydroformylated polyols when mixed with a crosslinker such as glycerol [56]. The effects of structural heterogeneity on the properties of hydroformylation-derived PUs have also been evaluated [53]. In one study, a commercial soybean oil-based hydroformylated polyol (hydroxyl number: 236 mg KOH/g) was partially esterified at different extents by formic acid to prepare a group of polyester polyols with hydroxyl numbers varying from 86 to 236 mg KOH/g [53]. PUs prepared from polyols with hydroxyl numbers larger than 200 mg KOH/g were glassy materials possessing high T_g and high crosslinking densities, and their properties were not negatively affected by the heterogeneity of polyol functionalities. In contrast, PUs obtained from polyols with hydroxyl numbers less than 200 mg KOH/g were rubbery materials with low T_g and cross-linking densities. The heterogeneity of polyol functionalities caused these rubbery PUs to have low strength and elongation, suggesting the necessity to consider polyol heterogeneity when developing flexible PUs [53]. In order to alleviate the negative effects of such heterogeneity on flexible PU applications, novel triols with high molecular weights and well-defined structure have been prepared from high-oleic sunflower oil via a series of modification and separation processes including methanolysis, fractionation, hydroformylation, hydrogenation, and polycondesation [55]. The resulting PU networks exhibited good elastomeric properties, indicating the suitability of triols for flexible PU foams applications [55]. The heterogeneity of hydroxyl functionality distribution of hydroformylated polyols has also been shown to negatively affect the performance of waterborne PU coatings [57]. Polyols with the narrowest functionality distribution resulted in coatings with the best balance of hardness, flexibility, and abrasion resistance, while polyols with the widest functionality distribution led to soft coatings with the lowest abrasion resistance.

2.2.3   Ozonolysis Pathway

Polyol production from vegetable oils by ozonolysis typically involves two steps (Scheme 2.4a): (1) formation of ozonide at the unsaturation sites of vegetable oils and simultaneous decomposition of ozonide into aldehyde and carboxylic acid; and (2) reduction of aldehyde into alcohols with a catalyst, such as Raney nickel. Because of the cleavage of all double bonds during ozonolysis, only one primary hydroxyl group is introduced at each unsaturated fatty acid chain no matter whether it is mono- or poly- unsaturated, thus polyols having a maximal functionality of three are obtained [58]. Depending on the composition of fatty acids of vegetable

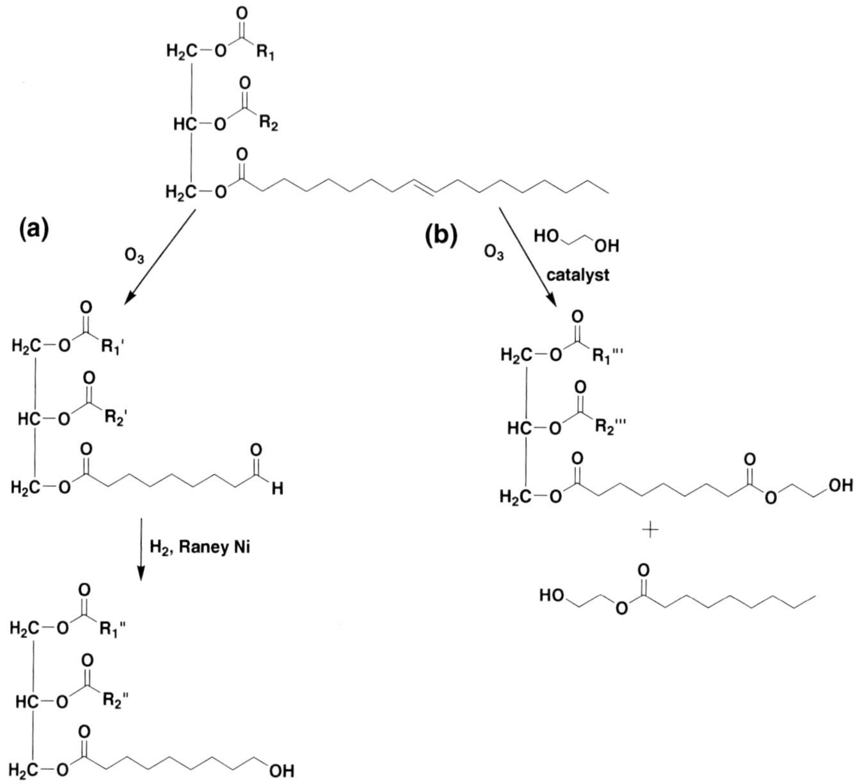

Scheme 2.4 Production of vegetable oil-based polyols via ozonolysis and hydrogenation pathway (a) and via ozonolysis pathway with the addition of ethylene glycol (b) (R_1 and R_2 are fatty acid side chains of vegetable oil; R_1', R_2', R_1'', R_2'', R_1''', and R_2''' are modified fatty acid side chains of vegetable oil)

oils, the ozonolysis-derived polyols are a mixture of different contents of mono-, di-, and tri-ols with triglyceride structures and saturated triglycerides. During the ozonolysis of vegetable oils, the cleavage of double bonds may also generate small alcohol molecules such as nonanol, 1, 3-propanediol, hexanol, and others [59], which could be valuable intermediates in chemical industries upon separation and purification. However, for the purpose of producing polyols for PU applications, small alcohol molecules are usually removed due to their detrimental effects on PU properties [58, 59]. Compared to polyols from epoxidization and ring-opening and from hydroformylation and hydrogenation, which have hydroxyl groups in the middle of fatty acids chains, ozonolysis-derived polyols only have terminal primary hydroxyl groups. As a result, ozonolysis-derived polyols have faster curing rates with isocyanates and effectively eliminate a majority of undesirable dangling chains. Ozonolysis-derived polyols typically have low molecular weights due to loss of part of the fatty acid chains that result from double bond cleavages [1]. Under different ozonolysis conditions, polyols prepared from canola oil have been reported to have varied properties, such as hydroxyl numbers from 152 to 260 mg KOH/g and acid numbers from 2 to 52 mg KOH/g [58–60]. The high acid numbers are attributed to carboxylic acids, which form during ozonation processes and cannot be reduced to alcohols under subsequent hydrogenation conditions [58–60].

Ozonolysis-derived polyols have also been produced via a one-step reaction without hydrogenation. Typically, multi-hydroxyl alcohols, such as ethylene glycol or glycerol, were mixed with vegetable oil and a catalyst (e.g., sodium hydroxide, calcium carbonate, and sulfuric acid). With the bubbling of ozone, the alcohol reacts with ozonide intermediates generated from ozonolysis of vegetable oil to form ester linkages, producing polyester polyols with terminal hydroxyl groups [61, 62]. Scheme 2.4b shows a schematic route for the production of polyester polyols by ozonolysis with the addition of ethylene glycol. Besides its simplicity and potential low cost, this one-step ozonolysis pathway has the advantage of producing polyols with a broad range of properties by using different multi-hydroxyl alcohols.

Due to their low contents of fatty acid dangling chains, ozonolysis-derived polyols produce highly crosslinked PUs that feature strong hydrogen bonding and superior mechanical properties such as compressive strength and Young's modulus [63, 64]. The properties of PUs from ozonolysis-derived polyols can be manipulated by factors such as polyol structures and NCO/OH molar ratios. PU produced from trilinolein-based polyols had a high T_g and showed properties typical of rigid plastics, such as a high tensile strength of 51 MPa and a low elongation of 25 %, while PU from soybean oil-based polyols had a low T_g, tensile strength of 31 MPa and an elongation of 176 %, showing hard rubber-like properties [58]. This difference of PU properties is mainly attributed to different functionalities of trilinolein-based polyols ($f = 3$) and soybean oil-based polyols ($f = 2.5$) and the presence of fatty acid dangling chains in soybean oil-based polyols. Compared to PUs prepared from epoxidized and hydroformylated polyols, PUs from ozonolysis-derived polyols with the same functionality exhibited improved mechanical properties and higher T_g due to the absence or lower content of fatty acid dangling chains [58]. When the NCO/OH molar ratio increased from 1.0 to 1.2, T_g of PU plastic sheets produced from ozonolysis-derived canola oil polyols increased from 23 to 43 °C [60]. An NCO/OH ratio of 1.2 or higher could result in the formation of imperfect elastic networks due to the decreased concentration of elastically active network chains in PU networks [60].

Zero or low volatile organic content (VOC) PU coatings from soybean oil glyceride polyols prepared by one-step ozonolysis with glycerol exhibited high hardness, gloss, and chemical resistance as well as excellent adhesion to metal surfaces [61]. They are suitable for industrial, automotive and architectural applications. Rigid PU foams prepared from a mixture of glycerol and the above soybean oil glyceride polyol at a 1:3 weight ratio also showed satisfactory mechanical and thermal properties, which were comparable to those of PU analogues from a commercial polyol [61]. The addition of glycerol not only increased the hydroxyl number of polyols but also acted as a crosslinker to improve the mechanical strength of the foams. Ozonolysis-derived polyols combined with poly(methyl methacrylate) (PMMA) have also been used to prepare sequential interpenetrating polymer networks (IPNs) with satisfactory mechanical properties. IPNs showed varying performance for different applications depending on the PMMA content in the PU structure [65]. Additionally, 100 % bio-based PUs have been prepared from ozonolysis-derived polyols from canola oil and two oleic acid-based isocyanates (1,16-diisocyanatohexadec-8-ene and 1,7-heptamethylene diisocyanate, Chap. 1) [66, 67]. Lower Young's modulus and higher elongation were observed in PU from the former isocyanate than from the latter one due to more flexibility of the long chain of 1,16-diisocyanatohexadec-8-ene [66]. With the same ozonolysis-derived polyols, PU derived from 1,7-heptamethylene diisocyanate showed properties comparable to those of an analogue from petroleum-based 1,6-hexamethylene diisocyanate [67].

2.2.4   Thiol-ene Coupling Pathway

Thiol-ene coupling reactions involve a free radical chain mechanism by which thiols are grafted onto double bonds. They are not sensitive to oxygen and can be carried out in the absence of photoinitiators through a photoreaction [68]. Due to its high conversion yield and fast reaction rate, a UV-initiated thiol-ene coupling reaction is generally used for preparing polyols from vegetable oils and their derivatives with 2-mercaptoethanol as a common thiol monomer (Scheme 2.5). Soybean oil-based polyols have also been prepared via heat-initiated thiol-ene coupling which required longer reaction time than UV-initiated thiol-ene coupling [69]. During the thiol-ene coupling process of vegetable oils and their derivatives, side reactions occurred, including disulfide formation, double bond isomerization, and inter- and intra-molecular bond formation [70]. Despite these side reactions, most byproducts contained hydroxyl functional groups and could participate in PU formation. Similar to polyols from epoxidation followed by oxirane ring-opening

Scheme 2.5 Production of vegetable oil-based polyols via UV-initiated thiol-ene coupling pathway using 2-mercaptoethanol as a thiol monomer (R_1 and R_2 are fatty acid side chains of vegetable oil; R_1' and R_2' are modified fatty acid side chains of vegetable oil)

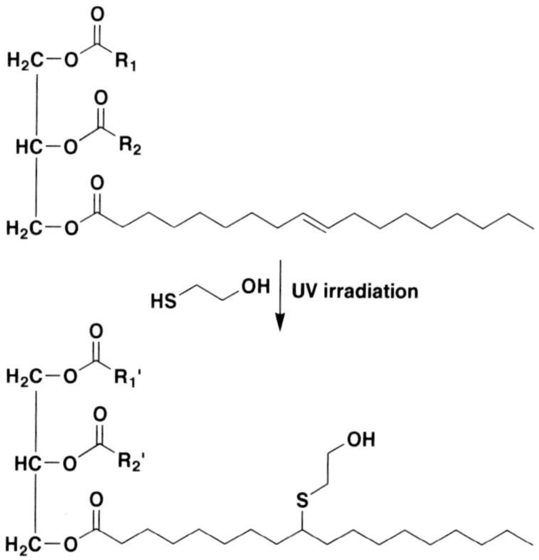

and from hydroformylation followed by hydrogenation, polyols from the thiol-ene coupling pathway also contain hydroxyl groups located in the middle of fatty acid chains, leaving part of the fatty acid chains as dangling components in PU structures. Currently, most thiol-ene coupling-derived polyols have been prepared from fatty acids or fatty acid esters. Reports on the preparation of vegetable oil-based polyols via thiol-ene coupling reactions are limited [69, 70]. Rapeseed oil-based polyols produced by UV-initiated thiol-ene coupling with 2-mercaptoethanol showed an acid number of 2.5 mg KOH/g, a hydroxyl number of 223 mg KOH/g, and an average functionality of 3.6. Soybean oil-based polyols produced by heat-initiated thiol-ene coupling with 2-mercaptoethanol showed an acid number of 2.5 mg KOH/g and a hydroxyl number of 200 mg KOH/g. PUs derived from these two polyols showed properties such as thermal and mechanical properties similar to those from a commercial polyol [69, 70].

2.2.5 Transesterification/Amidation Pathway

All of the above discussed pathways for the production of vegetable oil-based polyols take place at the double bond moieties of vegetable oils. Transesterification and amidation use a different approach that makes use of the ester moieties in the structures of vegetable oils to produce polyols (Scheme 2.6) [1]. Glycerol is the most predominantly used alcohol for the transesterification of vegetable oils, but the use of other alcohols, such as pentaerythritol [71] and triethanolamine [72], has also been reported. During transesterification, the addition of a small amount of soap acts as an emulsifier that can improve the compatibility between glycerol and triglycerides and thus increase the production efficiency of monoglycerides [73, 74]. Transesterification reactions are mostly catalyzed by organic and inorganic bases such as methoxides of sodium, calcium, and postassium [71, 75–77]; sodium hydroxide; and calcium hydroxide [73], and by metal oxides such as lead [78, 79] and calcium oxides [80]. Enzyme-catalyzed transesterification has also been reported [81]. Polyols produced from vegetable oils by transesterification with glycerol (i.e., glycerolysis) are a mixture of mono-, di-, and tri-glycerides and residual glycerol. Among these components, monoglycerides, which contain two hydroxyl groups per molecule, play an important role for PU production. Depending on reaction conditions and feedstocks used, polyols with monoglyceride contents ranging from 48.3 to 90.1 % and hydroxyl numbers ranging from 90 to 183 mg KOH/g have been obtained [73, 80, 82]. Since all hydroxyl groups in polyols derived from transesterification of vegetable oils, except for castor oil, are located on the glycerol backbone, all of the acid side chains would act as dangling components when polyols are crosslinked with isocyanates. In applications where flexibility is preferred, these dangling chains are beneficial due to their plasticizing effects, while in applications that require high rigidity, these plasticizing effects are detrimental and undesirable.

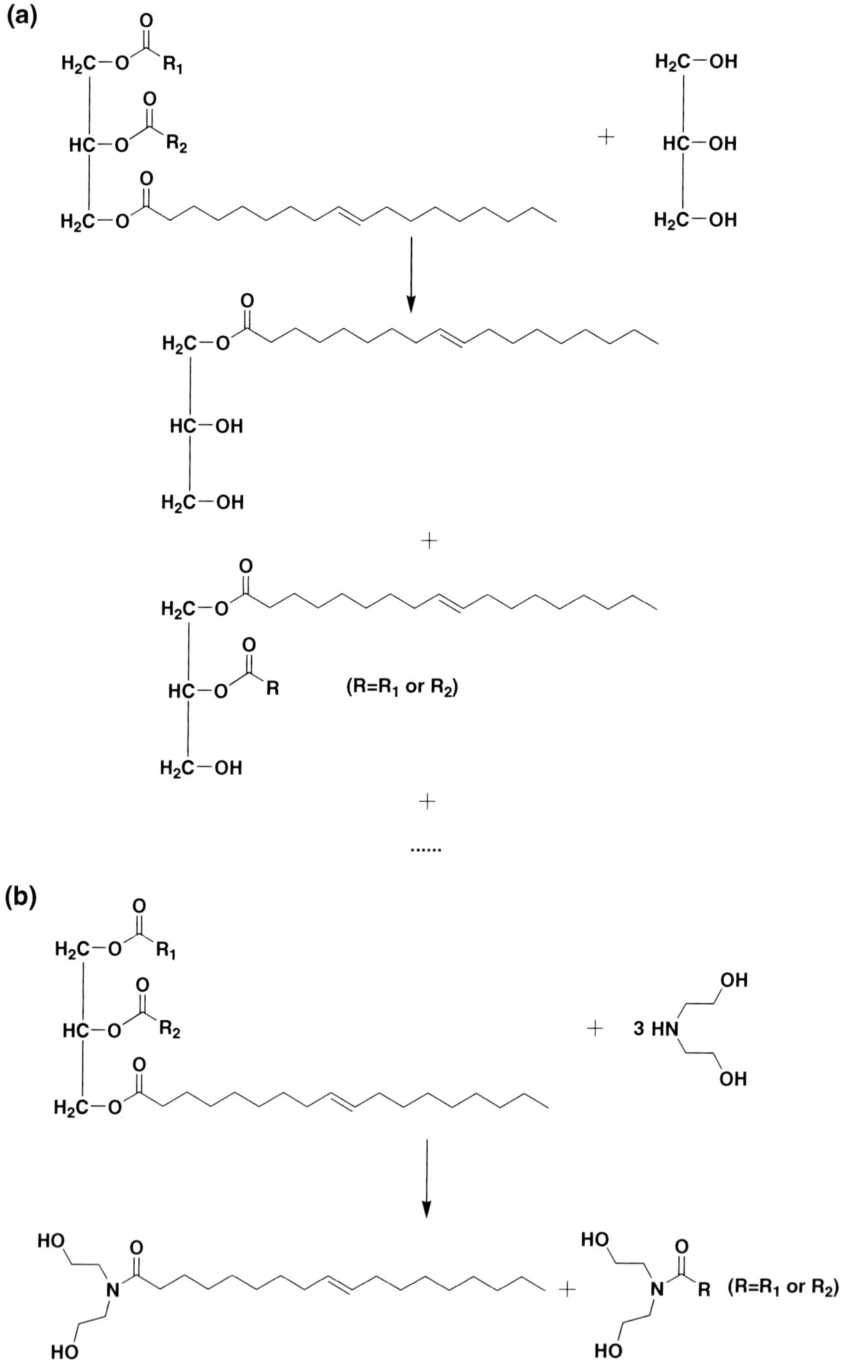

(a)

(b)

Scheme 2.6 Production of vegetable oil-based polyols via transesterification with glycerol (**a**) and amidation with diethanolamine (**b**) (R_1 and R_2 are fatty acid side chains of vegetable oil) [1]

Polyols produced by the transesterification of vegetable oils can be partially or completely substituted for petroleum-derived polyols for the preparation of PU foams and coatings. Compared to epoxidized and hydroxylated soy polyols, a mixture of glycerolysis-based soy polyols (90.1 % of monoglyceride, 1.3 % of diglyceride, and 8.6 % of glycerol) and glycerol propoxylate (Mw: 400 g/mol) exhibited higher reactivity with isocyanates and resulted in flexible PU foams with more uniform cell structure [73]. Increasing the portion of glycerolysis-based polyols from palm oil in blends with diethylene glycol (DEG)/polyethylene glycol (PEG, Mw: 200 g/mol) produced semi-rigid PU foams with higher flexibility. This was mainly caused by the increased content of monoglycerides, which served as soft segments in the foam structures [82]. A mixture of glycerolysis-based polyols from Nahar oil and PEG (Mw: 200 g/mol) has also been used to produce PU coatings [79]. As the molar ratio of NCO/OH increased, PU coatings showed improved properties including impact resistance, hardness, gloss, and adhesive strength, which can be attributed to the increased crosslinking densities in PU networks. By reacting with trimmers of isophorone diisocyanate, PU coatings based on 100 % glycerolysis-based polyols were prepared from linseed, soybean, and sesame oils. The produced coatings generally showed satisfactory flexibility and adhesion properties as well as good chemical resistance [83].

Similar to the above transesterification processes, amidation with amines, usually diethanolamine, can also convert vegetable oils into diethanol fatty acid amides for producing PU foams and coatings. Compared to commonly used transesterification with glycerol at 230–250 °C [73, 78, 80], the amidation of vegetable oils with diethanolamine are carried out at a lower temperature, usually at 110 °C [72, 84–87]. Amidation-derived polyols from vegetable oils, such as linseed, soybean, rapeseed, sunflower, coconut, Nahar, and cottonseed oils, have been used for the development of PU foams and coatings with satisfactory physical and mechanical properties [72, 84, 88–90]. PU resins from amidation-based polyols from Nahar and linseed oils showed superior coating performance, such as adhesion, gloss, hardness, and chemical resistance, compared to polyester resins from the same oils [84, 89].

2.3 Castor Oil-Based Polyols and Polyurethanes

As an exception to common vegetable oils, castor oil contains naturally occurring hydroxyl groups. Approximately 90 % of fatty acids in castor oil consist of ricinoleic acid, which is a mono-unsaturated 18-carbon fatty acid with a hydroxyl group on its 12th carbon [2] (Fig. 2.4). An in-depth structural analysis of castor oil showed that castor oil had an average hydroxyl functionality of 2.7, which resulted from the contributions of 70 % triols (triricinoleate of glycerol) and 30 % diols (triacylglycerols having only two ricinoleyl groups), and the absence of monoalcohols (triacylglycerols having one ricinoleyl group) [91]. Due to its naturally occurring hydroxyl groups and wide availability, castor oil has long been a versatile and valuable feedstock for direct use in the PU industry.

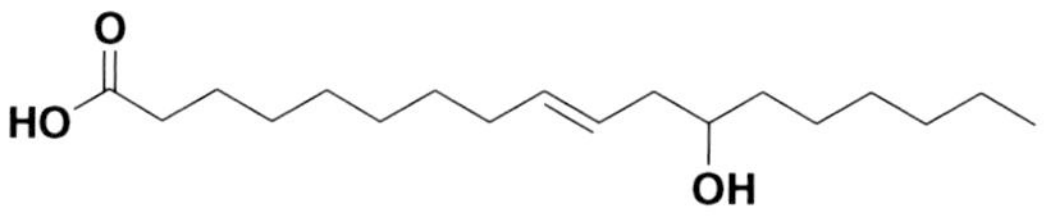

Ricinoleic acid

Fig. 2.4 Chemical structure of major fatty acid in castor oil

Castor oil has low functionality and possesses low reactivity due to the secondary hydroxyl groups in the ricinoleic acid chains, resulting in castor oil-based PUs with semi-flexible or semi-rigid properties [92]. Two major modification pathways have been widely used to improve the properties and applicability of castor oil-based polyols for producing PUs with improved properties and wider applications. One is the transesterification/amidation of castor oil using its ester moieties, and the other is the alkoxylation of castor oil using its hydroxyl groups. The functionality and hydroxyl number of castor oil-based polyols can be increased by transesterification with glycerol, pentaerythritol, and other polyols [1], or amidation with diethanolamine [93]. As a result, PUs with more rigid properties have been obtained. For example, castor oil-based polyols prepared by transesterification with triethanolamine and amidation with diethanolamine, showed hydroxyl numbers ranging from approximately 291 to 512 mg KOH/g. Their derived PU coatings exhibited higher tensile strengths (19.8–57.4 MPa) and glass transition temperatures (T_g, 44.5–84.5 °C) than those from an unmodified castor oil-based analogue (tensile strength: 14.1 MPa; T_g: 18.6 °C) [93]. Alkoxylation is a polymerization process by which epoxide monomers (e.g., ethylene oxide and propylene oxide) are incorporated into an alcohol for the formation of polyols. Castor oil can be converted to polyols with higher molecular weights and lower hydroxyl numbers by ethoxylation (alkoxylation with ethylene oxide) or propoxylation (alkoxylation with propylene oxide) [94]. Due to the incorporation of long polyether chains from castor oil, high quality flexible PUs such as foam mattresses can be produced. Scheme 2.7 shows a schematic synthetic route for castor oil-based polyols via ethoxylation.

In addition to these two modification methods, mixing castor oil with petrochemical-derived polyols is also an effective way to obtain fast reaction rates with isocyanates and to fine-tune product properties in the PU production process. By adding triisopropanolamine to the PU formulation, castor oil-based rigid PU foams showed increased compressive strength [95]. Millable PU elastomers prepared from castor oil and poly(propylene glycol) showed a wide range of physical and mechanical properties via varying PU formulations, such as the content of polyols and chain extenders [96]. They varied from soft elastomers to hard plastics. In comparison to a petroleum-based sample (e.g., Urepan 600), castor oil-based PU elastomers exhibited comparable tensile strength, compression set, and resilience and slightly inferior abrasion resistance and elongation at break [96]. Castor oil, in combination with recycled polyethylene terephthalate (PET), adipic acid, and polyethylene glycol (PEG), has been used to prepare PU coatings for insulation applications [97, 98]. Higher tensile strength and better electrical insulation

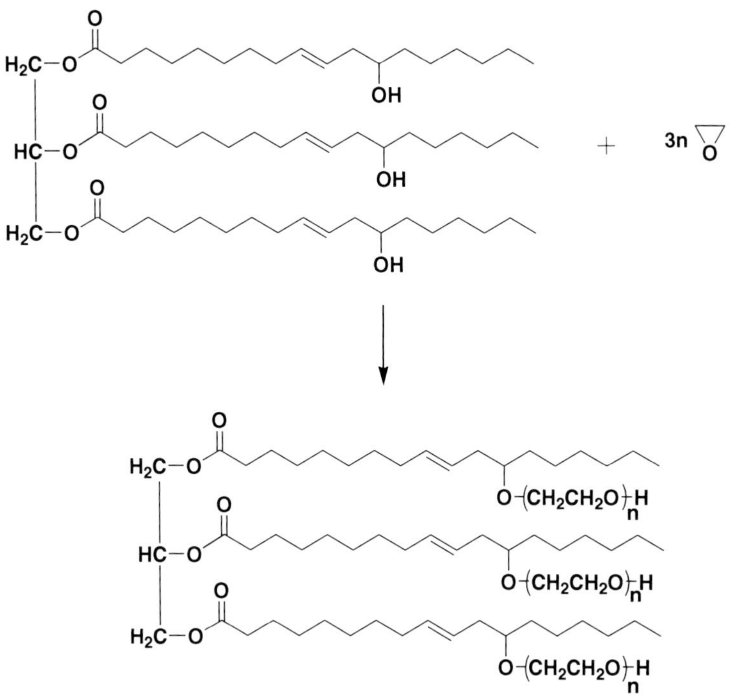

Scheme 2.7 Production of castor oil-based polyols via ethoxylation [94]

performance (comparable or superior to regular PU insulation) were generally obtained with increasing crosslinking density in PU networks. These coatings also exhibited excellent resistance to acid (1 or 10 % sulfuric acid) and alkaline (1 % sodium hydroxide) solutions, and a certain degree of swelling in toluene and DMF [97, 98]. Additionally, castor oil has been extensively used in the preparation of PU-based interpenetrating polymer networks (IPNs) [99–102]. PUs with 100 % bio-based materials have also been prepared from castor oil and soybean oil-based isocyanates, showing low Young's modulus and tensile strength due to the absence of hard segments [103].

2.4   Fatty Acid- and Fatty Acid Ester-Based Polyols and Polyurethanes

Fatty acids and fatty acid esters have carbon-carbon double bonds and carboxyl groups/ester linkages, which can be converted to hydroxyl groups with the above mentioned methods for vegetable oil-based polyols. Besides these methods [54, 67, 104–108], other methods have been reported for the synthesis of polyols from fatty

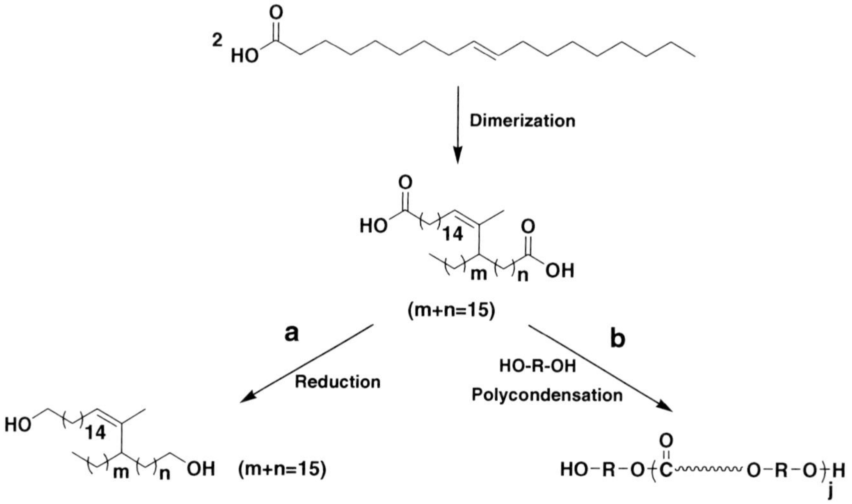

Scheme 2.8 Production of bio-based polyols via dimerization of oleic acid followed by reduction (**a**) or by polycondensation with glycols (**b**) [68, 110]

acids and fatty acid esters, including dimerization of fatty acids, cyclotrimerization of alkyne fatty acid esters, self-metathesis of fatty acids, cationic polymerization of epoxidized fatty acid esters, and hydrosilylation of fatty acid esters, all of which were followed by a reduction reaction.

Dimerization of fatty acids is a complex reaction that can proceed under various catalysts, such as alkaline metal salts, Lewis acids, and clays [109]. Through further reduction (Scheme 2.8a) or polycondensation with glycols (Scheme 2.8b), dimeric fatty acid diols/polyester polyols are obtained [68, 110]. Waterborne PU coatings synthesized from dimer fatty acid-based polyester polyols exhibited high water resistance and thermal stability but low toluene resistance and mechanical properties [110]. The introduction of adipic acid to dimer fatty acid-based polyester polyols could improve the toluene resistance and mechanical properties of dimer fatty acid-derived PU coatings.

Fatty acid-based cyclotrimerization (Scheme 2.9) has also been reported for producing polyols. Through a series of modifications, including bromation, dehydrobromination, and esterification, oleic acid and 10-undecenoic acid (derived from ricinoleic acid) were first converted to methyl 9-octadecynoate and methyl 10-undecynoate, which then underwent cyclotrimerization using heterogeneous Pd/C as a catalyst and subsequent reduction of ester groups with LiAlH$_4$ to yield primary hydroxyl groups [111]. Because of the plasticizing effect of the long aliphatic chains in oleic acid-derived aromatic triols, oleic acid-based PUs showed a lower T_g value than 10-undecenoic acid-based analogues.

Unsaturated linear diol with terminal primary hydroxyl groups was synthesized via the self-metathesis of oleic acid with a Grubbs catalyst followed by reduction with LiAlH$_4$ (Scheme 2.10) [105]. By reacting with an oleic acid-derived

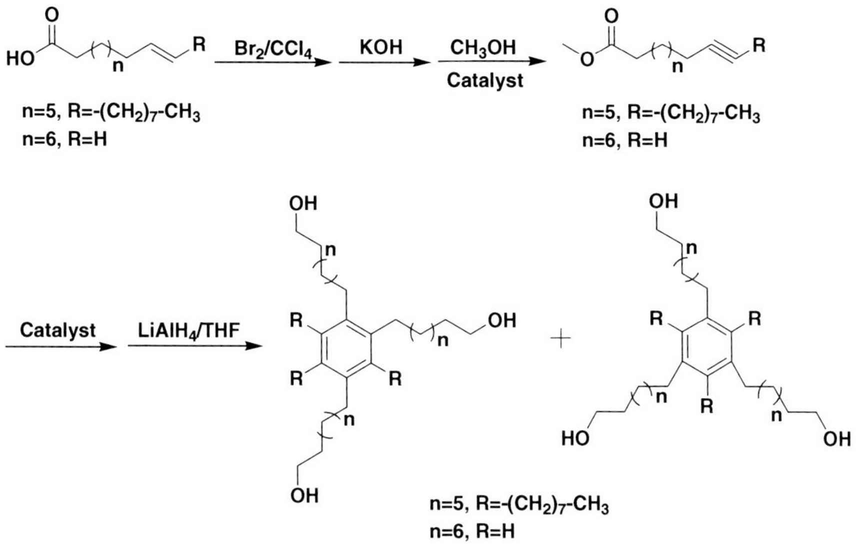

Scheme 2.9 Production of fatty acid-based polyols via cyclotrimerization pathway [111]

Scheme 2.10 Production of fatty acid-based linear diol via self-metathesis followed by reduction [105]

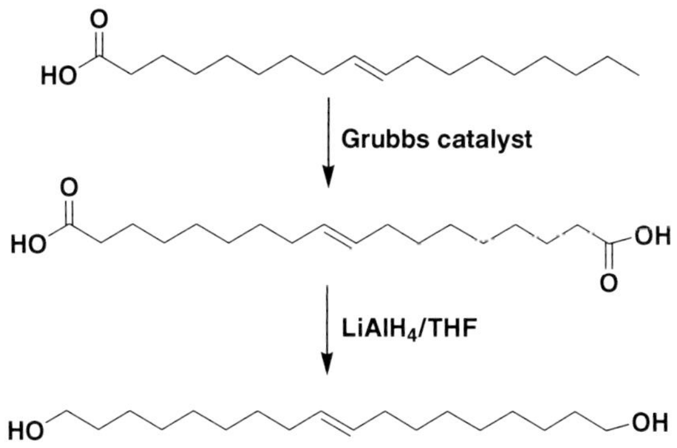

diisocyanate (1,7-heptamethylene diisocyanate, Chap. 1) in the presence of a bio-based chain extender (1,9-nonanediol, derived from ozonolyzed oleic acid) [67], 100 % bio-based thermoplastic PU (TPU) has been prepared from the unsaturated linear diol. The TPU showed similar phase behavior but lower tensile strength and elongation at break, compared to an analogue prepared from the diols with a petroleum-based diisocyanate (i.e., 1,6-hexamethylene diisocyanate). The difference of properties between these two TPUs was due to the effects of odd- and even-numbered methylene groups in 1,7-heptamethylene diisocyanate and 1,6-hexamethylene diisocyanate [105].

Oligomeric polyether polyols were synthesized via fluoroantimonic acid-catalyzed cationic polymerization of epoxidized methyl oleate and subsequent controlled reduction of ester groups (Scheme 2.11) [112]. Depending on the degree

Scheme 2.11 Production of methyl oleate-based polyether polyols [112]

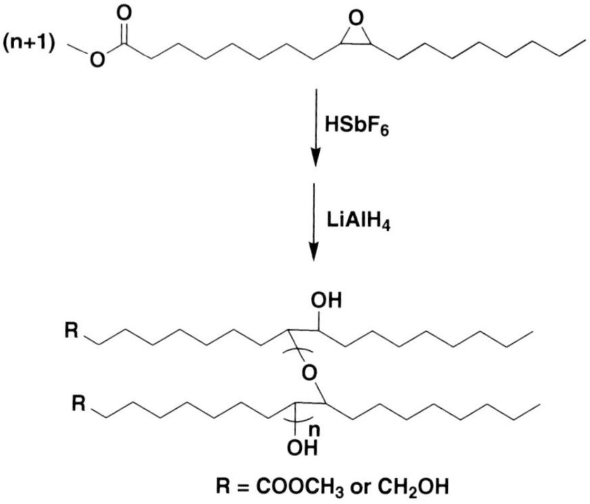

R = COOCH$_3$ or CH$_2$OH

of reduction reaction with lithium aluminum hydride, polyols had hydroxyl numbers ranging from 94 to 260 mg KOH/g and molecular weights from 1220 to 1149 g/mol, and varied from clear liquids to white, waxy solids at room temperature. Thermal and mechanical analyses indicated that the produced PUs could be used as hard rubber or rigid plastics [112].

Through platinum-catalyzed hydrosilylation with phenyltris(dimethylsiloxy) silane followed by reduction with LiAlH$_4$, methyl 10-undecenoate was converted to a silicon-containing polyol with terminal primary hydroxyl groups (Scheme 2.12), which had a hydroxyl number of 194 mg KOH/g [113]. The incorporation of silicone endowed the resulting PUs with enhanced thermal stability under atmospheric conditions, suggesting their potential applications as fire-retardant materials [113].

2.5 Crude Glycerol-Based Polyols and Polyurethanes

Recently, the feasibility of utilizing crude glycerol as a renewable feedstock for the production of polyols and PU foams and coatings has been investigated [9, 11, 12]. Through a one-pot thermochemical process, crude glycerol, in the presence of sulfuric acid, was successfully converted to polyols with suitable properties for applications of PU foams and coatings. The reactions involved in the thermochemical process mainly included the acidification of soap, esterification of glycerol and fatty acids, and transesterification of glycerol and fatty acid methyl esters, as shown in Scheme 2.13. Crude glycerol-based polyols were a mixture primarily consisting of monoglycerides, glycerol, and diglycerides. Under preferred reaction conditions of 200 °C, 90 min, and 3 % sulfuric acid loading, the crude glycerol-based polyols produced showed a hydroxyl number of approximately 481 mg KOH/g and an acid number of approximately 5 mg KOH/g. PU foams produced from this crude

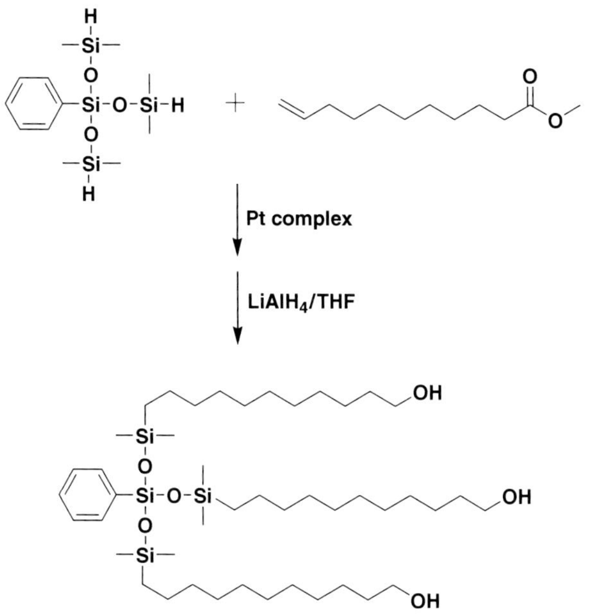

Scheme 2.12 Production of a silicone-containing polyol from fatty acid ester [113]

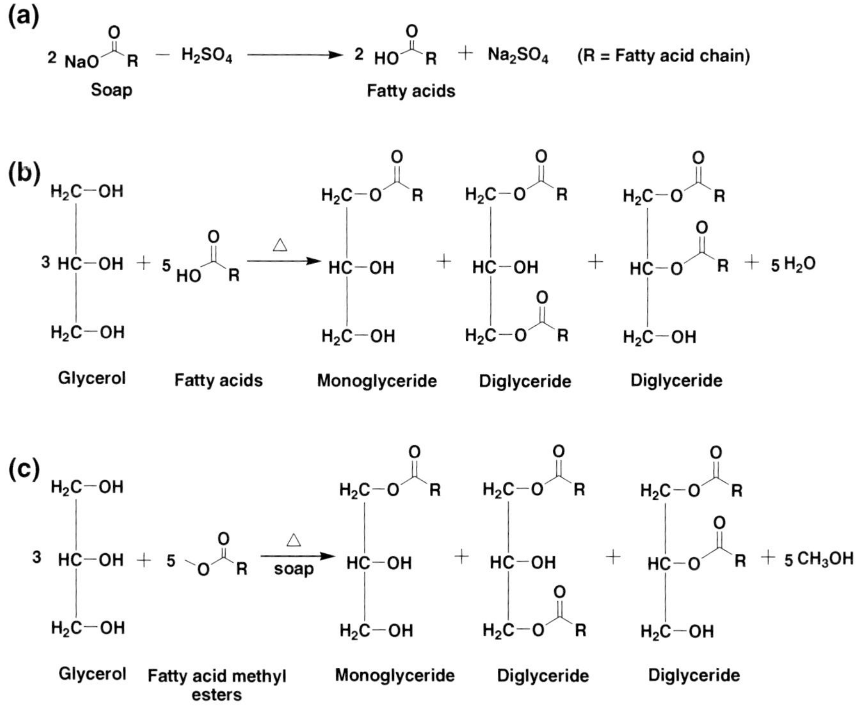

(a)

(b)

(c)

Scheme 2.13 Reactions mainly involved in the one-pot thermochemical conversion of crude glycerol for the production of polyols

glycerol-based polyol and polymeric methylene-4,4'-diphenyl diisocyanate (pMDI) presented a compressive strength of approximately 184.5 kPa and a density of approximately 43.0 kg/m^3, which were comparable to those of some petroleum-based analogs [9]. Crude glycerol has also been used for producing polyols and PU foams in combination with PET and diethylene glycol [10]. With an increase of the weight ratio of crude glycerol to PET and DEG, polyols showed increased hydroxyl numbers, which resulted in PU foams with increased density and compressive strength but decreased thermal stability. A decreased content of aromatic segments was responsible for the decreased thermal stability of PU foams. Under vacuum conditions, a crude glycerol-based polyol with a lower hydroxyl number (e.g., 378 mg KOH/g) was also prepared by a one-pot thermochemical process and had components similar to polyols obtained under atmospheric conditions [11]. Waterborne PU coatings from this polyol and isophorone diisocyanate (IPDI) showed excellent adhesion to steel surfaces, good pencil hardness, but relatively low flexibility. The incorporation of petroleum-based polyether polyols can improve the flexibility of crude glycerol-based waterborne PU coatings [11].

2.6 Summary and Future Prospects

Vegetable oils and their derivatives including fatty acids, fatty acid esters, and crude glycerol have good potential as renewable and sustainable feedstocks in producing bio-based polyols and PUs. Modifications made on the double bond and/or carbonyl moieties (i.e., ester linkages or carboxyl group) of vegetable oils and their derivatives allow the synthesis of polyols with different reactivities, functionalities, molecular weights, and other properties. Because of their high versatility, polyols from vegetable oils and their derivatives have been used to produce various PU materials such as foams, elastomers, rigid plastics, and coatings, which have shown properties mostly comparable to those of their petroleum-based analogs.

In spite of their promise, vegetable oils and their derivatives still face challenges such as technical and/or cost barriers to production of high resilient flexible foams. Future efforts in this field will be of high interest. Crude glycerol, a byproduct of the biodiesel industry, is a promising renewable feedstock for producing polyols and PUs. However, its varied composition makes it difficult to obtain polyols with consistent quality and properties. This problem will be effectively solved by adjusting the composition of crude glycerol with the addition of crude fatty acids or fatty acid methyl esters. The inherent structural heterogeneities of vegetable oils also challenge the production of polyols and PUs with consistent properties. The use of single components or derivatives, such as one type of fatty acid, and advances in genetic engineering should lead to polyols and PUs with more homogeneous structures and consistent qualities. Currently, bio-based polyols and PUs may still have higher costs than petroleum-based analogues. However, with the continuing advances in technologies and the inevitable depletion of the world's petroleum resources, the future of bio-based polyols and PUs looks very promising and bright.

References

1. Pfister DP, Xia Y, Larock RC (2011) Recent advances in vegetable oil-based polyurethanes. Chem Sus Chem 4:703–717
2. Petrović ZS (2008) Polyurethanes from vegetable oils. Polym Rev 48:109–155
3. USDA-FAS (2013) Production, supply and distribution online. http://apps.fas.usda.gov/psdonline/psdHome.aspx. Accessed Mar 2015
4. Babb DA (2012) Polyurethanes from renewable resources. In: Synthetic biodegradable polymers. Springer, Berlin
5. Johnson DT, Taconi KA (2007) The glycerin glut: options for the value-added conversion of crude glycerol resulting from biodiesel production. Environ Prog 26:338–348
6. Hu S, Luo X, Wan C, Li Y (2012) Characterization of crude glycerol from biodiesel plants. J Agr Food Chem 60:5915–5921
7. Pachauri N, He B (2006) Value-added utilization of crude glycerol from biodiesel production: a survey of current research activities. In: Proceedings of the ASABE annual international meeting 9
8. Hu S, Wan C, Li Y (2012) Production and characterization of biopolyols and polyurethane foams from crude glycerol based liquefaction of soybean straw. Bioresour Technol 103:227–233
9. Luo X, Hu S, Zhang X, Li Y (2013) Thermochemical conversion of crude glycerol to biopolyols for the production of polyurethane foams. Bioresour Technol 139:323–329
10. Luo X, Li Y (2014) Synthesis and characterization of polyols and polyurethane foams from PET waste and crude glycerol. J Polym Environ 22:318–328
11. Hu S, Luo X, Li Y (2014) Production of polyols and waterborne polyurethane dispersions from biodiesel-derived crude glycerol. J Appl Polym Sci. doi:10.1002/app.41425
12. Li C, Luo X, Li T, Tong X, Li Y (2014) Polyurethane foams based on crude glycerol-derived biopolyols: one-pot preparation of biopolyols with branched fatty acid ester chains and its effects on foam formation and properties. Polymer 55:6529–6538
13. Pan X, Webster DC (2012) New biobased high functionality polyols and their use in polyurethane coatings. ChemSusChem 5:419–429
14. Sinadinović-Fišer S, Janković M, Petrović ZS (2001) Kinetics of in situ epoxidation of soybean oil in bulk catalyzed by ion exchange resin. J Am Oil Chem Soc 78:725–731
15. Petrović ZS, Zlatanić A, Lava CC, Sinadinović-Fišer S (2002) Epoxidation of soybean oil in toluene with peroxoacetic and peroxoformic acids-kinetics and side reactions. Eur J Lipid Sci Tech 104:293–299
16. Kong X, Liu G, Curtis JM (2012) Novel polyurethane produced from canola oil based poly (ether ester) polyols: synthesis, characterization and properties. Eur Polym J 48:2097–2106
17. Cai C, Dai H, Chen R, Su C, Xu X, Zhang S, Yang L (2008) Studies on the kinetics of in situ epoxidation of vegetable oils. Eur J Lipid Sci Tech 110:341–346
18. Vlček T, Petrović ZS (2006) Optimization of the chemoenzymatic epoxidation of soybean oil. J Am Oil Chem Soc 83:247–252
19. Miao S, Zhang S, Su Z, Wang P (2010) A novel vegetable oil-lactate hybrid monomer for synthesis of high-Tg polyurethanes. J Polym Sci Pol Chem 48:243–250
20. Rüsch gen Klaas M, Warwel S (1999) Complete and partial epoxidation of plant oils by lipase-catalyzed perhydrolysis. Ind Crop Prod 9:125–132
21. Rüsch gen Klaas M, Warwel S (1996) Chemoenzymatic epoxidation of unsaturated fatty acid esters and plant oils. J Am Oil Chem Soc 73:1453–1457
22. Dai H, Yang L, Lin B, Wang C, Shi G (2009) Synthesis and characterization of the different soy-based polyols by ring-opening of epoxidized soybean oil with methanol, 1, 2-ethanediol and 1,2-propanediol. J Am Oil Chem Soc 86:261–267
23. Guo A, Cho Y, Petrović ZS (2000) Structure and properties of halogenated and nonhalogenated soy-based polyols. J Polym Sci Polym Chem 38:3900–3910

24. Kiatsimkul PP, Suppes GJ, Hsieh Fh, Lozada Z, Tu YC (2008) Preparation of high hydroxyl equivalent weight polyols from vegetable oils. Ind Crop Prod 27:257–264
25. Wang CS, Yang LT, Ni BL, Shi G (2009) Polyurethane networks from different soy-based polyols by the ring-opening of epoxidized soybean oil with methanol, glycol, and 1,2-propanediol. J Appl Polym Sci 114:125–131
26. Caillol S, Desroches M, Boutevin G, Loubat C, Auvergne R, Boutevin B (2012) Synthesis of new polyester polyols from epoxidized vegetable oils and biobased acids. Eur J Lipid Sci Tech 114:1447–1459
27. Monteavaro LL, da Silva EO, Costa AP, Samios D, Gerbase AE, Petzhold CL (2005) Polyurethane networks from formiated soy polyols: synthesis and mechanical characterization. J Am Oil Chem Soc 82:365–371
28. Sharmin E, Ashraf SM, Ahmad S (2007) Synthesis, characterization, antibacterial and corrosion protective properties of epoxies, epoxy-polyols and epoxy-polyurethane coatings from linseed and *Pongamia glabra* seed oils. Int Biol Macromol 40:407–422
29. Hu YH, Gao Y, Wang DN, Hu CP, Zu S, Vanoverloop L, Randall D (2002) Rigid polyurethane foam prepared from a rape seed oil based polyol. J Appl Polym Sci 84:591–597
30. Zlatanić A, Lava C, Zhang W, Petrović ZS (2004) Effect of structure on properties of polyols and polyurethanes based on different vegetable oils. J Polym Sci Pol Phys 42:809–819
31. Ionescu M, Petrović ZS, Wan X (2007) Ethoxylated soybean polyols for polyurethanes. J Polym Environ 15:237–243
32. Guo Y, Hardesty JH, Mannari VM, Massingill JL Jr (2007) Hydrolysis of epoxidized soybean oil in the presence of phosphoric acid. J Am Oil Chem Soc 84:929–935
33. Miao S, Zhang S, Su Z, Wang P (2013) Synthesis of bio-based polyurethanes from epoxidized soybean oil and isopropanolamine. J Appl Polym Sci 127:1929–1936
34. Petrović ZS, Yang L, Zlatanić A, Zhang W, Javni I (2007) Network structure and properties of polyurethanes from soybean oil. J Appl Polym Sci 105:2717–2727
35. Wang C, Yang L, Ni B, Wang L (2009) Thermal and mechanical properties of cast polyurethane resin based on soybean oil. J Appl Polym Sci 112:1122–1127
36. Petrović ZS, Guo A, Zhang W (2000) Structure and properties of polyurethanes based on halogenated and nonhalogenated soy-polyols. J Polym Sci Polym Chem 38:4062–4069
37. Petrović ZS, Zhang W, Zlatanic A, Lava CC, Ilavskyý M (2002) Effect of OH/NCO molar ratio on properties of soy-based polyurethane networks. J Polym Environ 10:5–12
38. Guo A, Javni I, Petrović Z (2000) Rigid polyurethane foams based on soybean oil. J Appl Polym Sci 77:467–473
39. Tan S, Abraham T, Ference D, Macosko CW (2011) Rigid polyurethane foams from a soybean oil-based Polyol. Polymer 52:2840–2846
40. Tu Y-C, Kiatsimkul P, Suppes G, Hsieh F-H (2007) Physical properties of water-blown rigid polyurethane foams from vegetable oil-based polyols. J Appl Polym Sci 105:453–459
41. Das S, Dave M, Wilkes GL (2009) Characterization of flexible polyurethane foams based on soybean-based polyols. J Appl Polym Sci 112:299–308
42. Pawlik H, Prociak A (2012) Influence of palm oil-based polyol on the properties of flexible polyurethane foams. J Polym Environ 20:438–445
43. Zhang L, Jeon HK, Malsam J, Herrington R, Macosko CW (2007) Substituting soybean oil-based polyol into polyurethane flexible foams. Polymer 48:6656–6667
44. Rojek P, Prociak A (2012) Effect of different rapeseed-oil-based polyols on mechanical properties of flexible polyurethane foams. J Appl Polym Sci 125:2936–2945
45. Prociak A, Rojek P, Pawlik H (2012) Flexible polyurethane foams modified with natural oil based polyols. J Cell Plast 48:489–499
46. Lu Y, Larock RC (2008) Soybean-oil-based waterborne polyurethane dispersions: effects of polyol functionality and hard segment content on properties. Biomacromolecules 9:3332–3340
47. Lu Y, Larock RC (2007) New hybrid latexes from a soybean oil-based waterborne polyurethane and acrylics via emulsion polymerization. Biomacromolecules 8:3108–3114

48. Lu Y, Larock RC (2010) Aqueous cationic polyurethane dispersions from vegetable oils. ChemSusChem 3:329–333
49. Lu Y, Larock RC (2010) Soybean oil-based, aqueous cationic polyurethane dispersions: synthesis and properties. Prog Org Coat 69:31–37
50. Lu Y, Larock RC (2011) Synthesis and properties of grafted latices from a soybean oil-based waterborne polyurethane and acrylics. J Appl Polym Sci 119:3305–3314
51. Lu Y, Xia Y, Larock RC (2011) Surfactant-free core-shell hybrid latexes from soybean oil-based waterborne polyurethanes and poly(styrene-butyl acrylate). Prog Org Coat 71:336–342
52. Guo A, Demydov D, Zhang W, Petrović ZS (2002) Polyols and polyurethanes from hydroformylation of soybean oil. J Polym Environ 10:49–52
53. Petrović ZS, Guo A, Javni I, Cvetković I, Hong DP (2008) Polyurethane networks from polyols obtained by hydroformylation of soybean oil. Polym Int 57:275–281
54. Petrović ZS, Cvetković I, Milic J, Hong D, Javni I (2012) Hyperbranched polyols from hydroformylated methyl soyate. J Appl Polym Sci 125:2920–2928
55. Petrović ZS, Cvetković I, Hong D, Wan X, Zhang W, Abraham TW, Malsam J (2010) Vegetable oil-based triols from hydroformylated fatty acids and polyurethane elastomers. Eur J Lipid Sci Tech 112:97–102
56. Guo A, Zhang W, Petrović ZS (2006) Structure-property relationships in polyurethanes derived from soybean oil. J Mater Sci 41:4914–4920
57. Argyropoulos J, Popa P, Spilman G, Bhattacharjee D, Koonce W (2009) Seed oil based polyester polyols for coatings. J Coat Technol Res 6:501–508
58. Petrović ZS, Zhang W, Javni I (2005) Structure and properties of polyurethanes prepared from triglyceride polyols by ozonolysis. Biomacromolecules 6:713–719
59. Narine SS, Yue J, Kong X (2007) Production of polyols from canola oil and their chemical identification and physical properties. J Am Oil Chem Soc 84:173–179
60. Kong X, Narine SS (2007) Physical properties of polyurethane plastic sheets produced from polyols from canola oil. Biomacromolecules 8:2203–2209
61. Benecke HP, Vijayendran BR, Garbark DB, Mitchell KP (2008) Low cost and highly reactive biobased polyols: a co-product of the emerging biorefinery economy. Clean-Soil Air Water 36:694–699
62. Tran P, Graiver D, Narayan R (2005) Ozone-mediated polyol synthesis from soybean oil. J Am Oil Chem Soc 82:653–659
63. Narine SS, Kong X, Bouzidi L, Sporns P (2007) Physical properties of polyurethanes produced from polyols from seed oils: I. Elastomers. J Am Oil Chem Soc 84:55–63
64. Narine SS, Kong X, Bouzidi L, Sporns P (2007) Physical properties of polyurethanes produced from polyols from seed oils: II. Foams. J Am Oil Chem Soc 84:65–72
65. Kong X, Narine SS (2008) Sequential interpenetrating polymer networks produced from vegetable oil based polyurethane and poly(methyl methacrylate). Biomacromolecules 9:2221–2229
66. Hojabri L, Kong X, Narine SS (2010) Novel long chain unsaturated diisocyanate from fatty acid: synthesis, characterization, and application in bio-based polyurethane. J Polym Sci Pol Chem 48:3302–3310
67. Hojabri L, Kong X, Narine SS (2009) Fatty acid-derived diisocyanate and biobased polyurethane produced from vegetable oil: synthesis, polymerization, and characterization. Biomacromolecules 10:884–891
68. Desroches M, Escouvois M, Auvergne R, Caillol S, Boutevin B (2012) From vegetable oils to polyurethanes: synthetic routes to polyols and main industrial products. Polym Rev 52:38–79
69. Caillol S, Desroches M, Carlotti S, Auvergne R, Boutevin B (2012) Synthesis of new polyurethanes from vegetable oil by thiol-ene coupling. Green Mater 1:16–26
70. Desroches M, Caillol S, Lapinte V, Auvergne R, Boutevin B (2011) Synthesis of biobased polyols by thiol-ene coupling from vegetable oils. Macromolecules 44:2489–2500
71. Chuayjuljit S, Maungchareon A, Saravari O (2010) Preparation and properties of palm oil-based rigid polyurethane nanocomposite foams. J Reinf Plast Comp 29:218–225

72. Stirna U, Cabulis U, Beverte I (2008) Water-blown polyisocyanurate foams from vegetable oil polyols. J Cell Plast 44:139–160
73. Campanella A, Bonnaillie LM, Wool RP (2009) Polyurethane foams from soyoil-based polyols. J Appl Polym Sci 112:2567–2578
74. Can E, Küsefoglu S, Wool RP (2001) Rigid, thermosetting liquid molding resins from renewable resources. I. Synthesis and polymerization of soy oil monoglyceride maleates. J Appl Polym Sci 81:69–77
75. Yunus R, Fakhru'I-Razi A, Ooi TL, Biak DRA, Iyuke SE (2004) Kinetics of transesterification of palm-based methyl esters with trimethylolpropane. J Am Oil Chem Soc 81:497–503
76. Petrović ZS, Cvetković I, Hong D, Wan X, Zhang W, Abraham T, Malsam J (2008) Polyester polyols and polyurethanes from ricinoleic acid. J Appl Polym Sci 108:1184–1190
77. Gryglewicz S, Piechocki W, Gryglewicz G (2003) Preparation of polyol esters based on vegetable and animal fats. Bioresour Technol 87:35–39
78. Dutta N, Karak N, Dolui SK (2004) Synthesis and characterization of polyester resins based on Nahar seed oil. Prog Org Coat 49:146–152
79. Dutta S, Karak N (2006) Effect of the NCO/OH ratio on the properties of *Mesua Ferrea L.* seed oil-modified polyurethane resins. Polym Int 55:49–56
80. Bakare IO, Pavithran C, Okieimen FE, Pillai CKS (2008) Synthesis and characterization of rubber-seed-oil-based polyurethanes. J Appl Polym Sci 109:3292–3301
81. Bhabhe MD, Athawale VD (1998) Chemoenzymatic synthesis of urethane oil based on special functional group oil. J Appl Polym Sci 69:1451–1458
82. Tanaka R, Hirose S, Hatakeyama H (2008) Preparation and characterization of polyurethane foams using a palm oil-based polyol. Bioresour Technol 99:3810–3816
83. Gite VV, Kulkarni RD, Hundiwale DG, Kapadi UR (2006) Synthesis and characterisation of polyurethane coatings based on trimer of isophorone diisocyanate (IPDI) and monoglycerides of oils. Surf Coat Int Pt B-C 89:117–122
84. Dutta S, Karak N (2005) Synthesis, characterization of poly (urethane amide) resins from Nahar seed oil for surface coating applications. Prog Org Coat 53:147–152
85. Lee CS, Ooi TL, Chuah CH, Ahmad S (2007) Synthesis of palm oil-based diethanolamides. J Am Oil Chem Soc 84:945–952
86. Palanisamy A, Karuna MSL, Satyavani T, Kumar DR (2011) Development and characterization of water-blown polyurethane foams from diethanolamides of karanja oil. J Am Oil Chem Soc 88:541–549
87. Palanisamy A, Rao BS, Mehazabeen S (2011) Diethanolamides of castor oil as polyols for the development of water-blown polyurethane foam. J Polym Environ 19:698–705
88. Khoe TH, Frankel EN (1976) Rigid polyurethane foams from diethanolamides of carboxylated oils and fatty acids. J Am Oil Chem Soc 53:17–19
89. Yadav S, Zafar F, Hasnat A, Ahmad S (2009) Poly(urethane fatty amide) resin from linseed oil-a renewable resource. Prog Org Coat 64:27–32
90. Meshram PD, Puri RG, Patil AL, Gite VV (2013) High performance moisture cured poly (ether-urethane) amide coatings based on renewable resource (cottonseed oil). J Coat Technol Res 10:331–338
91. Trân NB, Vialle J, Pham QT (1997) Castor oil-based polyurethanes: 1. Structural characterization of castor oil-nature of intact glycerides and distribution of hydroxyl groups. Polymer 38:2467–2473
92. Mutlu H, Meier MA (2010) Castor oil as a renewable resource for the chemical industry. Eur J Lipid Sci Tech 112:10–30
93. Stirna U, Lazdina B, Vilsone D, Lopez MJ, del Vargas-Garcia Carmen M, Suárez-Estrella F, Moreno J (2012) Structure and properties of the polyurethane and polyurethane foam synthesized from castor oil polyols. J Cell Plast 48:476–488
94. Corcuera MA, Rueda L, Fernandez dArlas B, Arbelaiz A, Marieta C, Mondragon I, Eceiza A (2010) Microstructure and properties of polyurethanes derived from castor oil. Polym Degrad Stabil 95:2175–2184

95. Lyon CK, Garrett VH, Goldblatt LA (1962) Solvent-blown, rigid urethane foams from low cost castor oil-polyol mixtures. J Am Oil Chem Soc 39:69–71
96. Yeganeh H, Mehdizadeh MR (2004) Synthesis and properties of isocyanate curable millable polyurethane elastomers based on castor oil as a renewable resource polyol. Eur Polym J 40:1233–1238
97. Yeganeh H, Shamekhi MA (2006) Novel polyurethane insulating coatings based on polyhydroxyl compounds, derived from glycolysed PET and castor oil. J Appl Polym Sci 99:1222–1233
98. Yeganeh H, Moeini HR (2007) Novel polyurethane electrical insulator coatings based on amide-ester-ether polyols derived from castor oil and re-cycled poly(ethylene terphthalate). High Perform Polym 19:113–126
99. Athawale V, Kolekar S (1998) Interpenetrating polymer networks based on polyol modified castor oil polyurethane and polymethyl methacrylate. Eur Polym J 34:1447–1451
100. Das SK, Lenka S (2000) Interpenetrating polymer networks composed of castor oil-based polyurethane and 2-hydroxy-4-methacryloyloxy acetophenone. J Appl Polym Sci 75:1487–1492
101. Xie HQ, Guo JS (2002) Room temperature synthesis and mechanical properties of two kinds of elastomeric interpenetrating polymer networks based on castor oil. Eur Polym J 38:2271–2277
102. Chen S, Wang Q, Wang T, Pei X (2011) Preparation, damping and thermal properties of potassium titanate whiskers filled castor oil-based polyurethane/epoxy interpenetrating polymer network composites. Mater Des 32:803–807
103. Çayli G, Küsefoğlu S (2008) Biobased polyisocyanates from plant oil triglycerides: synthesis, polymerization, and charaterization. J Appl Polym Sci 109:2948–2955
104. Junming X, Jianchun J, Jing L (2012) Preparation of polyester polyols from unsaturated fatty acid. J Appl Polym Sci 126:1377–1384
105. Hojabri L, Kong X, Narine SS (2010) Functional thermoplastics from linear diols and diisocyanates produced entirely from renewable lipid sources. Biomacromolecules 11:911–918
106. González-Paz RJ, Lluch C, Lligadas G, Ronda JC, Galià M, Cádiz V (2011) A green approach toward oleic- and undecylenic acid-derived polyurethanes. J Polym Sci Pol Chem 49:2407–2416
107. Desroches M, Cuillol S, Auvergne R, Boutevin B (2012) Synthesis of pseudo-telechelic diols by transesterification and thiol-ene coupling. Eur J Lipid Sci Tech 114:84–91
108. Palaskar DV, Boyer A, Cloutet E, Le Meins J-F, Gadenne B, Alfos C, Farcet C, Cramail H (2012) Original diols from sunflower and ricin oils: synthesis, characterization, and use as polyurethane building blocks. J Polym Sci Pol Chem 50:1766–1782
109. Tolvanen P, Mäki-Arvela P, Kumar N, Eränen K, Sjöholm R, Hemming J, Holmbom B, Salmi T, Murzin DY (2007) Thermal and catalytic oligomerisation of fatty acids. Appl Cata A-Gen 330:1–11
110. Liu X, Xu K, Liu H, Cai H, Su J, Fu Z, Guo Y, Chen M (2011) Preparation and properties of waterborne polyurethanes with natural dimer fatty acids based polyester polyol as soft segment. Prog Org Coat 72:612–620
111. Lligadas G, Ronda JC, Galià M, Cádiz V (2007) Polyurethane networks from fatty-acid-based aromatic triols: synthesis and characterization. Biomacromolecules 8:1858–1864
112. Lligadas G, Ronda JC, Galià M, Biermann U, Metzger JO (2006) Synthesis and characterization of polyurethanes from epoxidized methyl oleate based polyether polyols as renewable resources. J Polym Sci Pol Chem 44:634–645
113. Lligadas G, Ronda JC, Galià M, Cádiz V (2006) Novel silicon-containing polyurethanes from vegetable oils as renewable resources synthesis and properties. Biomacromolecules 7:2420–2426

Chapter 3
Lignocellulosic Biomass-Based Polyols for Polyurethane Applications

Abstract Recently, there has been increased interest in developing bio-based polyols and polyurethanes (PUs) from lignocellulosic biomass. As the world's most abundant renewable feedstock, lignocellulosic biomass is rich in hydroxyl groups and has potential as a feedstock to produce bio-based polyols for PU applications. Lignocellulosic biomass can be converted to liquid polyols through oxypropylation or liquefaction processes. The produced liquid polyols from lignocellulosic biomass can be used to prepare various PU products, such as foams, films, and adhesives. The properties of lignocellulosic biomass-derived polyols and PUs depend upon various factors, such as feedstock characteristics, reaction parameters, and PU formulations. The major challenge for lignocellulosic biomass-based polyols is the consumption of solvents, which comprise a large fraction of the reactants, during liquefaction processes.

Keywords Lignocellulosic biomass · Oxypropylation · Liquefaction · Bio-based polyols · Polyurethanes

3.1 Introduction

Lignocellulosic biomass, such as forest waste and crop residues, is considered to be the world's most abundant renewable biomass feedstock. It is an inexpensive feedstock for the development of biofuels and bioproducts in the chemical industry [1] and has attracted considerable attention in the polyurethane (PU) industry. The main structural units of lignocellulosic biomass consist of cellulose, hemicellulose, and lignin, all of which are highly functional and have a large number of hydroxyl groups. Cellulose is a linear polymer of β-glucoses connected via multiple β-C1–O–C4′ bonds (Fig. 3.1). Hemicellulose is a branched heteropolymer containing both pentose and hexose units, such as xylose, arabinose, glucose, mannose, and galactose. Xylan is a major component of hemicellulose and has diverse structures depending on the botanic source. The general structure of xylan is a linear backbone

© The Author(s) 2015   45
Y. Li et al., *Bio-based Polyols and Polyurethanes*,
SpringerBriefs in Green Chemistry for Sustainability,
DOI 10.1007/978-3-319-21539-6_3

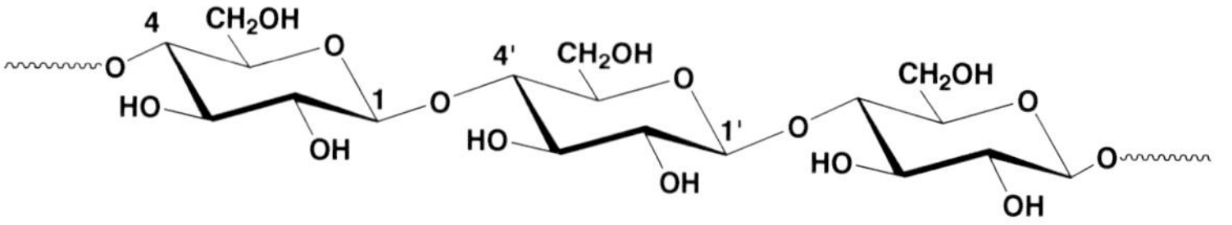

Fig. 3.1 Chemical structure of cellulose

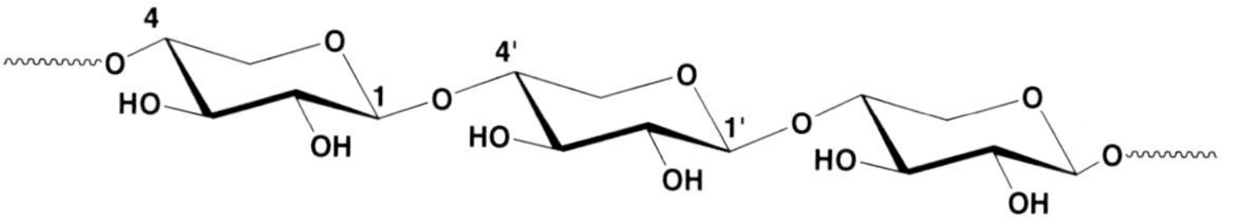

Fig. 3.2 Chemical structure of xylan

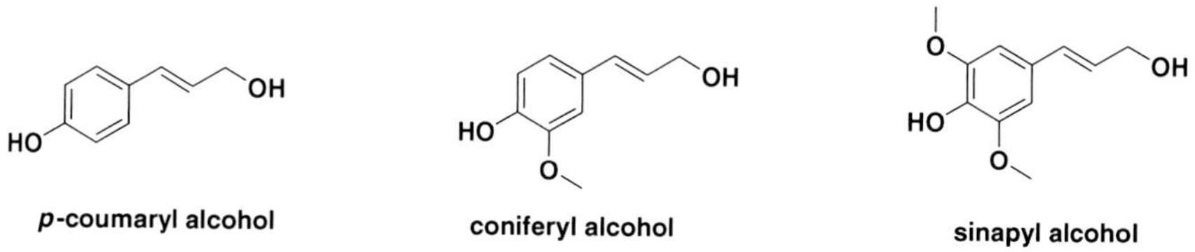

p-coumaryl alcohol   coniferyl alcohol   sinapyl alcohol

Fig. 3.3 Three major lignin-building monomers: *p*-coumaryl alcohol, coniferyl alcohol, and sinapyl alcohol

consisting of D-xylopyranose units via β-C1–O–C4′ linkages (Fig. 3.2). Lignin is a highly branched aromatic polymer with various phenylpropane-derived units that are formed from three major monomers, i.e., *p*-coumaryl alcohol, coniferyl alcohol, and sinapyl alcohol (Fig. 3.3).

A variety of lignocellulosic biomass including wood, crop residues, food processing wastes, and byproducts of biorefineries, have been investigated for the production of polyols and PUs. Table 3.1 summarizes the chemical composition of some lignocellulosic biomass feedstocks. Due to its solid nature, chemical modification is necessary to convert lignocellulosic biomass into liquid polyols for the production of PUs. Although lignocellulosic biomass can be used directly to produce PU products in combination with polyols [2], limited access and low reactivity of functional groups of the biomass are encountered because of steric constraints. In this process, the PU formulations usually contain less than 30 % lignocellulosic biomass, which acts as a filler in the PUs [3–5]. The incorporation of less than 30 % lignocellulosic biomass has been shown to improve Young's modulus and crosslink density of the PUs compared to those prepared from polyether polyols only, while PUs with more than 30 % lignocellulosic biomass were rigid and brittle [3]. Thus, lignocellulosic biomass is normally converted into polyols which are then used to produce PUs.

Since 2010, the pulp and paper industry has generated approximately 55 million tons per year of lignin-rich residue most of which is used as a source of

Table 3.1 Chemical composition of lignocellulosic biomass

Lignocellulosic biomass	Cellulose (%)	Hemicellulose (%)	Lignin (%)	Ash (%)	References
Wheat straw	40.5	29.9	17.2	9.8	[6]
Corn stover	34.2	23.9	6.6	N/A[a]	[7]
Corn cob	37.0	30.0	23.0	2.0	[1]
Bamboo residue	47.5	18.8	23.7	1.6	[8]
Sugarcane bagasse	49.2	21.7	20.1	1.6	[9]
Sugar beet pulp	23.0	31.0	4.0	N/A[a]	[10]
Date seeds	20.0	55.0	23.0	1.1	[1]
Olive stone	34.0	25.0	25.0	1.0	[1]
Apple pomace	21.0	23.0	19.0	1.7	[1]
Rapeseed cake	18.0	44.0	25.0	1.0	[1]

Note [a]Not available

Table 3.2 Characteristics of kraft lignin, soda lignin, and organosolv lignin [14, 15]

Lignins	Number average molecular weight (M_n)	Weight average molecular weight (M_w)	Polydispersity index (PDI)	Average hydroxyl functionality
Kraft lignin	1220	11,880	9.7	10
Soda lignin	1100	6800	6.2	9
Organosolv lignin	700	1700	2.4	6

energy via combustion [11, 12]. Currently, only about 2 % of lignin-rich residue is utilized for the production of lignin-based chemicals and materials [12]. Several processing methods have been developed to isolate lignin and, depending on the process, lignin products with different properties are obtained [13]. Among them, kraft lignin, soda lignin, and organosolv lignin are three important ones, which have been widely studied for the production of PUs. Kraft lignin is obtained from softwood by the Kraft-pulping process. It is a highly hydrophobic polymer with aliphatic thiol groups. Soda lignin is obtained by the soda pulping process, which is usually used to process non-woody biomass, such as bagasse and wheat straw [13]. Because no sulfur-containing chemicals are added in the soda pulping process, soda lignin is sulfur-free. Organosolv lignin is obtained by the organosolv pulping process in which organic solvents are used to isolate lignin from wood or other lignocellulosic biomass, especially hardwood. The solvents commonly used in this process include methanol, ethanol, formic acid, and acetic acid [13]. Organosolv lignin is also a sulfur-free polymer and has a lower molecular weight and hydroxyl functionality compared to kraft lignin and soda lignin (Table 3.2), which make organosolv lignin a more desired feedstock for polyol production (see Sect. 3.2.2.1).

Currently, polyols from lignocellulosic biomass are mainly produced through two major approaches: (1) oxypropylation and (2) liquefaction. This chapter focuses on the mechanisms and processes of these two methods for polyol production from lignocellulosic biomass. The properties of polyols and the resulting PUs and the factors that affect their properties are also reviewed.

3.2 Oxypropylation of Lignocellulosic Biomass

3.2.1 Mechanism

The oxypropylation of lignocellulosic biomass is a polymerization process that grafts propylene oxide onto macromolecular structures of lignocellulosic biomass. In general, two approaches are used for oxypropylation of biomass to produce polyols. One is to directly mix biomass, propylene oxide, and a catalyst (usually KOH in the form of commercial pellets). The other is to activate the biomass first in an ethanol solution of KOH for several hours at room temperature under nitrogen atmosphere, and then dry the activated substrate by the removal of ethanol, followed by mixing with propylene oxide. Usually, this process is conducted under high pressure (ca. 650–1820 kPa) and high temperature (ca. 100–200 °C) for simultaneous or sequential activation and oxypropylation [16, 17]. For the sequential approach, biomass is first activated by the catalyst, followed by oxypropylation reaction with propylene oxide. Scheme 3.1 shows a schematic representation of the oxypropylation reaction of lignocellulosic biomass. Most hydroxyl groups of lignocellulosic biomass, especially phenolic ones from the lignin component, are actually wrapped inside the molecule [2]. During oxypropylation, propylene oxide chains are introduced and the hydroxyl groups of lignocellulosic biomass are moved to the polyether chain ends to form polyols. The number of hydroxyl groups in oxypropylated lignocellulosic biomass is the same as that in originally unmodified lignocellulosic biomass.

Oxypropylation-derived polyols generally are mixtures of different compounds, including oxypropylated lignocellulosic biomass, poly(propylene oxide) homopolymers, and some unreacted or partially oxypropylated lignocellulosic biomass, all of which can be used directly for the preparation of PUs [17]. Oxypropylated lignocellulosic biomass is a branched polymer with many hydroxyl groups in each macromolecule. In general, a much higher viscosity is obtained for oxypropylated

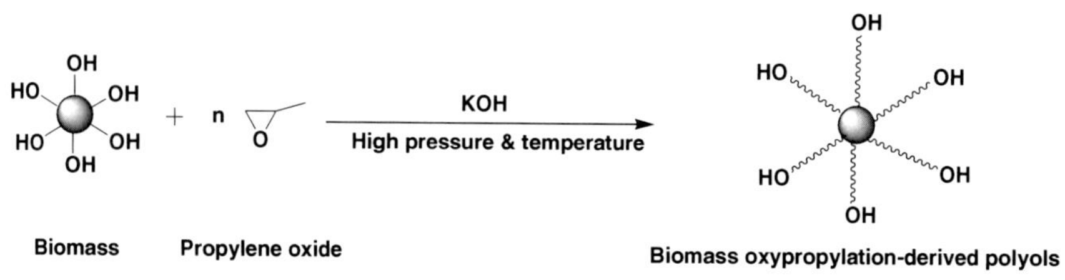

Scheme 3.1 Oxypropylation reaction of lignocellulosic biomass

lignocellulosic biomass than for poly(propylene oxide) homopolymers, which have two terminal hydroxyl groups in oxypropylation-derived polyols, and the viscosity varies by controlling the oxypropylation process and the molecular weight of the grafted poly(propylene oxide) chain [18]. Because of the low viscosity and linear structure, the presence of poly(propylene oxide) homopolymers in oxypropylation-derived polyols can act as a diluent to decrease the viscosity of the polyols as well as a chain transfer agent during reaction with isocyanates for the production of PUs. Depending on the characteristics of lignocellulosic biomass and oxypropylation parameters, such as the reaction temperature, catalyst loading, and ratio of propylene oxide to lignocellulosic biomass, polyols with different factions of oxypropylated biomass/poly(propylene oxide) homopolymers are obtained [19]. Solvents, such as hexane, can selectively separate poly(propylene oxide) homopolymers from oxypropylated lignocellulosic biomass [18].

3.2.2 Process and Parameters

Polyols from oxypropylation of lignocellulosic biomass are usually prepared using a high-pressure and high-temperature stainless steel reactor equipped with a mixer and temperature and pressure controllers. The oxypropylation reaction occurs with an initial progressive increase in pressure and temperature, followed by a dramatic decrease of the pressure to a constant low level, indicating the completion of the oxypropylation reaction. After cooling, oxypropylation-derived polyols are obtained as a viscous liquid for PU applications.

Fourier transform-infrared spectroscopy (FT-IR) and/or ^{1}H nuclear magnetic resonance (^{1}H NMR) are usually used to further assess the occurrence of oxypropylation of biomass [2, 14, 16, 20, 21]. In a typical FT-IR spectrum of oxypropylation-derived polyols, several features are observed: (a) increased absorption peaks at 2800–2970 cm^{-1} due to the C–H stretching of aliphatic CH, CH$_2$, and CH$_3$ groups; (b) increased absorption peak at 1000–1100 cm^{-1} due to the stretching of C–O bonds of ether moieties; and (c) the appearance of an absorption peak at 1375 cm^{-1} due to the C–H bending of CH$_3$ groups from propylene oxide units [2, 20, 21]. These features indicate the successful grafting of poly(propylene oxide) onto biomass macromolecules. The presence of peaks at 1.1–1.2 and 3.0–4.0 ppm due to the protons of CH$_3$ and CH$_2$ and of CH groups of propylene oxide units in ^{1}H NMR spectra provide further evidence of the occurrence of the oxypropylation of biomass feedstock [14, 20, 21].

3.2.2.1 Effect of Lignocellulosic Biomass Feedstock

The molecular weight of the lignin, which varies with the source of lignin and processing method (Table 3.2), has been shown to affect reaction efficiency and polyol properties. A study on the oxypropylation of kraft lignin and organosolv lignin showed that organosolv lignin reacted more easily with propylene oxide than kraft lignin [14]. The oxypropylation of organosolv lignin was completed within 1.5 h, while the oxypropylation of kraft lignin required approximately 10 h. This result was due to the lower molecular weight of organosolv lignin. Compared to kraft lignin-derived polyols, organosolv lignin-derived polyols had lower poly (propylene oxide) homopolymer content. No solid residue was observed in organosolv lignin-derived polyols, while there was approximately 33–41 % solid residue present in kraft lignin-derived polyols [5].

Particle size of lignocellulosic biomass also has significant impacts on the reaction efficiency and the properties of the resulting polyols. As the particle size of sugar beet pulp increased from 0.5 mm (ground pulp) to 1 cm (pulp pellets), the solid residue content of the polyols increased from 0 to 60 %, while the viscosity of polyols decreased from 49 to 4 Pa.s [10]. Studies have shown that only the superficial hydroxyl groups participated in oxypropylation, thus decreased particle size will result in higher reactivity of the lignocellulosic biomass, lower solid residue content in the final products, and higher viscosity of the polyols. Usually, biomass is ground to less than 2 mm before being converted to value-added products [16, 20, 22].

3.2.2.2 Effect of Catalyst

KOH is the most commonly used catalyst in the oxypropylation of lignocellulosic biomass due to its ability to obtain high conversion efficiency [2, 17]. Besides KOH, other base catalysts such as NaOH, triethylamine, and 1, 4-diazabicyclo[2.2.2]octane have also been tested for oxypropylation of lignocellulosic biomass [10]. Among these four catalysts, both KOH and NaOH showed higher catalytic activities than the two tertiary amines, i.e., triethylamine, and 1, 4-diazabicyclo[2.2.2]octane. Compared to NaOH, less reaction time was required to complete the oxyproplyation process when catalyzed by KOH [10, 19]. When KOH was used as the catalyst, increased catalyst loading was shown to accelerate the oxypropylation of lignocellulosic biomass [10, 19] and increase the viscosity of the resulting polyols [2, 10]. As KOH loading increased further, homopolymerization of propylene oxide was promoted, leading to polyols with lower viscosities attributable to the low viscosity of poly(propylene oxide) homopolymers [10]. An optimum catalyst loading was suggested.

3.2.2.3 Effect of Propylene Oxide-to-Biomass Ratio

The propylene oxide-to-biomass ratio (v/w) is a critical factor affecting the performance of biomass oxypropylation. Under a fixed catalyst loading, oxypropylation-derived liquid polyols tended to show increased viscosity and hydroxyl number, but decreased molecular weight as the fraction of biomass increased in the reactor [2, 14]. These results were attributed to the formation of shorter grafting chains on the biomass structures. In contrast, with a higher propylene oxide-to-biomass ratio, polyols had lower viscosities and higher molecular weights, suggesting the formation of longer grafting chains on biomass structures [2]. The propylene oxide-to-biomass ratio also had an influence on the content of solid residues present in the final product. As the ratio of kraft lignin or sugar beet pulp to propylene oxide increased, the residual solid content in the oxypropylation-derived liquid polyols increased [14].

3.2.3 Properties of Oxypropylation-Derived Polyols

To assess their suitability in PU applications, polyols are typically characterized in terms of hydroxyl number, acid number, and viscosity. Very few highly acidic compounds are formed in oxypropylation-derived polyols; therefore, acid numbers are seldom measured or reported in studies on polyols from oxypropylation of lignocellulosic biomass. As the presence of poly(propylene oxide) homopolymers has a significant impact on the properties of oxypropylation-derived polyols, its content is usually measured as one of the important polyol properties. The properties of polyols also vary with the lignocellulosic biomass feedstock and oxypropylation parameters, such as type and loading of catalyst and ratio of propylene oxide to lignocellulosic biomass [17].

The properties of polyols produced by the oxypropylation of lignocellulosic biomass, such as wood and byproducts of biorefineries, are summarized in Table 3.3. Depending on the specific oxypropylation parameters and biomass type, oxypropylation-derived polyols showed hydroxyl numbers ranging from approximately 82 to 610 mg KOH/g, viscosities from 2.3 to 2860 Pa.s, residue contents from 0 to 41 %, and poly(propylene oxide) homopolymer contents from 2 to 83 %. Generally, oxypropylation-derived polyols are suitable for the production of rigid PU

Table 3.3 General properties of oxypropylation-derived polyols from lignocellulosic biomass

Lignocellulosic biomass	Polyol properties				References
	Hydroxyl number (mg KOH/g)	Viscosity (Pa.s)	Homopolymer content (wt%)	Solid residue content (wt%)	
Kraft lignin	148–180	2.3–4.2	75–83	33–41	[14]
Organosolv lignin	198–305	5.2–2860	2–42	0	[2, 14]
Soda lignin	82–120	15.2–165	18–30	0	[14]
Sugar beet pulp	260–418	49–710	N/A[a]	0–10	[10, 19]
Rapeseed cake	610	84.2	N/A[a]	N/A[a]	[21]
Cork	~131	2.9–13.0	68.7–75.8	0.2–4.9	[15, 16]

Note [a]Not available

foams, but PU products with higher flexibility can also be produced from oxypropylation-derived polyols with lower hydroxyl numbers and longer poly(propylene oxide) grafting chains [2]. Additionally, partial oxypropylation of biomass has been used for the development of thermoplastic composite materials [15, 23].

3.2.4 Polyurethanes Produced from Polyols Derived from Oxypropylation

Oxypropylation-derived polyols are most commonly used for the production of rigid PU foams through the reaction with aromatic isocyanates, such as toluene diisocyanate (TDI), methylene diphenyl diisocyanate (MDI), and polymeric MDI (pMDI), in the presence of amine-containing catalysts, water, and surfactants. Rigid PU foams produced from polyols derived from oxypropylation of organosolv lignin and soda lignin showed good insulating properties with thermal conductivities ranging from 0.015 to 0.024 W/(m.K) [14]. Thermal conductivities increased slightly under both natural and accelerated aging tests (i.e., aging tests at room temperature for 3 months and at 70 °C for 10 days), which suggested good resistance against natural and accelerated aging for rigid PU foams from these polyols. The insulation properties and dimensional stability of these rigid PU foams were comparable to rigid PU analogs prepared from petroleum-based polyols [14]. However, rigid PU foams from polyols derived from oxypropylation of kraft lignin showed higher thermal conductivity than those derived from organosolv lignin and soda lignin. Deformation was observed after its aging at room temperature for 3 months and at 70 °C for 10 days [14]. Rigid PU foams from polyols derived from oxypropylation of cork (hydroxyl number: 131 mg KOH/g, viscosity: 10.8 Pa.s) had a density of 28.4 kg/m^3 and thermal conductivity of 0.038 W/(m.K) [16]. When the polyols were blended with petroleum-derived polyols, the thermal conductivity of the resulting rigid PU foams decreased from 0.038 to 0.031 W/(m.K) as the content of the petroleum-derived polyols increased from 0 to 79 % in the polyol mixture [16].

3.3 Liquefaction of Lignocellulosic Biomass

3.3.1 Mechanism

Liquefaction of biomass is a process in which biomass is degraded and decomposed into smaller molecules by polyhydric alcohols via solvolytic reactions. Liquefaction of lignocellulosic biomass can be either acid- or base-catalyzed with the former being more common. Generally, the liquefaction of hemicellulose, lignin, and amorphous cellulose occurs rapidly during the early stages of the liquefaction process because they have amorphous structures that are easily accessible to liquefaction solvents.

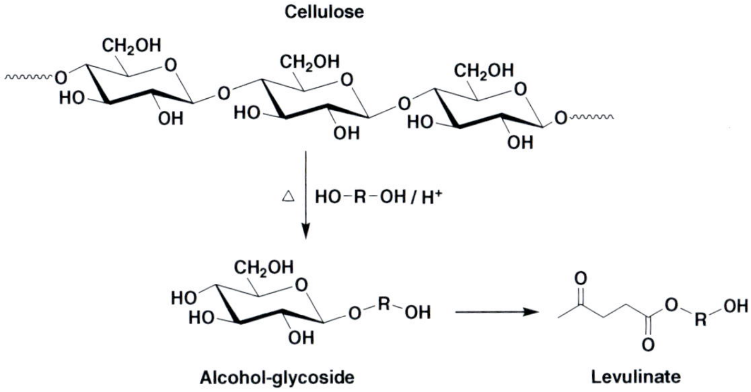

Scheme 3.2 Reaction mechanism of acid-catalyzed liquefaction of cellulose in polyhydric alcohols

In contrast, the liquefaction of crystalline cellulose is typically slower and continues through the later stages of the liquefaction process, because it has a well-packed molecular structure that is less accessible to the solvents [6, 24–26]. For this reason, cellulose liquefaction is commonly considered to be the rate-limiting step in the biomass liquefaction process [27, 28]. Scheme 3.2 shows one of the major lique-faction reactions occurring during the acid-catalyzed liquefaction of cellulose. The cellulose is first decomposed by solvolytic reactions into glucose or other small cellulose derivatives that can react with the liquefaction solvent to form glycoside derivatives. Then, the produced glycoside derivatives undergo further reactions to form levulinic acid and/or levulinates.

Biomass liquefaction is a complicated process in which a large number of chemical reactions occur and compete against each other simultaneously [29, 30]. Recondensation reactions among biomass derivatives and/or liquefaction solvents are one of the reactions which compete against other liquefaction reactions during the process. When dominant, these recondensation reactions can decrease the liq-uefaction efficiency by increasing the percentage of insoluble residues in liquefaction-derived polyols. The negative effects of recondensation reactions on liquefaction can be largely mitigated or avoided by optimization of liquefaction parameters, such as the process temperature and time, catalyst loading, and biomass-to-solvent ratio. The mechanism of recondensation reactions in acid-catalyzed biomass liquefaction processes was studied by liquefying cellulose, steamed white birch wood chips, alkali lignin, and their mixtures under the same liquefaction conditions [27]. No recondensation reactions occurred when cellulose or lignin was liquefied alone, even at a prolonged liquefaction time of 480 min. In contrast, when cellulose and lignin were liquefied together for 480 min, significant recondensation reactions were observed, and the insoluble residue content reached 50 and 76 % for cellulose/steamed wood and cellulose/alkali lignin mixtures, respectively. Based on these observations, it was suggested that recondensation reactions that occurred during acid-catalyzed liquefaction processes were largely caused by reactions between depolymerized cellulose and degraded aromatic derivatives from lignin and/or by the nucleophilic displacement reaction of cellulose by phenoxide ions [27].

3.3.2 Process and Parameters

The liquefaction process is usually conducted at elevated temperatures (150–250 °C) under atmospheric pressure using polyhydric alcohols, such as polyethylene glycol (PEG) and glycerol, as liquefaction solvents [30]. Microwave heating has been reported to achieve faster liquefaction of lignocellulosic biomass than traditional heating due to the direct conversion of electromagnetic energy into heat at the molecular level [31].

Liquefaction-derived polyols commonly contain solid residues in varying amounts depending on the liquefaction efficiency. Thus, the optimization of liquefaction parameters is usually conducted to obtain polyols with low residue content for the production of PUs. Factors such as the feedstock characteristics, liquefaction solvent, catalyst, and liquefaction temperature/time have significant effects on the efficiency of the liquefaction of lignocellulosic biomass.

3.3.2.1 Effect of Lignocellulosic Biomass Feedstock

The efficiency of liquefaction varies with the composition, structure, and morphology of the lignocellulosic biomass feedstock. A study on the liquefaction of different species of hardwood and softwood showed that four softwood species (*Sugi*, *Karamatsu*, *Akamatsu*, and *Radiata pine*) had significantly different liquefaction behaviors, while three hardwood species (*Udaikanba*, *Buna*, and *Mizunara*) shared similar liquefaction behaviors [32]. Compared to the hardwoods, the softwoods exhibited faster liquefaction rates but had earlier occurrences of unfavorable recondensation reactions due to the existence of large amounts of guaiacyl propane units in the softwoods, which were more reactive than the syringyl propane units in the hardwoods. Another study on the liquefaction of three types of waste paper (i.e., box paper, newspaper, and business paper) and wood (*Betula maximomawiczii regel*) found that the liquefaction rate varied, from highest to lowest, for wood, newspaper, business paper, and box paper [25]. The higher liquefaction efficiencies of wood and newspaper were explained by their higher contents of lignin and hemicellulose, which can be liquefied rapidly at the early stages of the process. Different liquefaction behaviors of crop residues, were also observed through comparison of feedstocks such as cotton stalks and bagasse [9], and corn stover, rice straw, and wheat straw [33]. During liquefaction with sulfuric acid as the catalyst, bagasse showed relatively higher liquefaction efficiency than cotton stalks. Corn stover had higher liquefaction efficiency compared to rice straw and wheat straw, both of which presented similar liquefaction efficiencies.

3.3.2.2 Effect of Solvent

Liquefaction solvents play a paramount role during the liquefaction process. The commonly used liquefaction solvents are polyhydric alcohols including glycerol, polyethylene glycol (PEG), and ethylene glycol (EG), which not only promote rapid and effective biomass liquefaction but also possess suitable polyol features for PU production. Most liquefaction processes are carried out at high liquefaction solvent-to-biomass weight ratios (approximately 3:1 to 5:1) to achieve high liquefaction efficiency. The biomass liquefaction-derived polyols are composed of liquefied biomass and a residue of liquefaction solvent or its derivatives, which constitute a major portion of the polyols due to the large volumes of liquefaction solvent used. These residues have significant impacts on polyol properties. For example, a binary mixture of PEG400 (MW: 400 g/mol) and glycerol is a commonly used liquefaction solvent to produce polyols with properties suitable for rigid or semi-rigid PU foam applications. In contrast, PEG4000 with an average molecular weight of 4000 g/mol was used as a sole liquefaction solvent to produce polyols suitable for highly resilient PU foam applications [30].

In general, the PEG400 to glycerol weight ratio at 4:1 showed high liquefaction efficiency and ability to dampen detrimental recondensation reactions [6, 24, 25, 34, 35]. However, a 4:1 mixture of PEG400/glycerol is not optimal for all types of biomass because of the structurally heterogeneous feature of lignocellulosic biomass. Studies have suggested different PEG400/glycerol weight ratios for different biomass feedstocks, such as 9:1 for some wood species, bagasse, and cotton stalks [9, 32] and 2:3 for acid hydrolysis residues of corn cobs [26].

Besides polyhydric alcohols, cyclic carbonates, such as ethylene carbonate (EC) and propylene carbonate (PC) derived from ethylene glycol and propylene glycol, respectively, have also been used for the rapid liquefaction of lignocellulosic biomass. Almost complete liquefaction (i.e., less than 2 % solid residues) of cellulose was achieved within 40 min at 150 °C for EC and PC with a solvent-to-biomass weight ratio of 5:1, while incomplete liquefaction was observed after 120 min for PEG400/glycerol (w/w: 4:1) under the same condition [36]. The high liquefaction efficiency of cyclic carbonates was explained by their high permittivity that leads to the high acid potential of liquefaction solvents [36]. However, in a later study on the liquefaction of corn stover (CS) using EC as a liquefaction solvent at different solvent-to-biomass weight ratios (i.e., EC/CS = 5:1 to 5:2), higher contents of solid residues, ranging from 2.0 to 11.6 %, were observed after 90 min of liquefaction at 170 °C [7]. These different observations might have been caused by the different compositions and morphologies of biomass feedstocks used in the liquefaction processes.

Currently, almost all liquefaction solvents are petroleum-derived and relatively expensive. To reduce the high cost and increase the renewability of the biomass liquefaction processes, the use of crude glycerol, a byproduct from biodiesel production processes, was recently reported as an alternative solvent to liquefy soybean straw for polyol production [22]. Despite its relatively low biomass liquefaction efficiency (around 70 % under optimal reaction conditions), the polyols produced from the liquefaction of soybean straw in crude glycerol and the resulting PU foams exhibited properties comparable to those obtained from conventional petrochemical solvent-based liquefaction processes. More importantly, this study indicated that certain impurities in crude glycerol, such as FFAs, soap and FAMEs, improved the properties of biomass liquefaction-derived polyols and PU foams due to their synergistic interactions with glycerol and/or biomass components [37]. These results suggested that crude glycerol can be directly used as a liquefaction solvent for polyol production from biomass without expensive pretreatments, while most of the current value-added conversion processes for crude glycerol require removal of impurities before processing. A direct use of crude glycerol to liquefy lignocellu-losic biomass for polyol production could potentially improve the economics and sustainability of liquefaction processes.

3.3.2.3 Effect of Catalyst

The liquefaction of lignocellulosic biomass can be conducted under either acid or base catalysis, of which the former is more commonly used. Among all acids investigated so far, concentrated (98 %) sulfuric acid has the highest catalytic ability in biomass liquefaction [38]. Generally, significant improvement on biomass liq-uefaction efficiency was observed by increasing sulfuric acid loading from 1 to 3 % based on the weight of the liquefaction solvent [9, 26, 38]. A further increase of sulfuric acid loading provided little improvement on biomass liquefaction effi-ciency, but increased the risk of recondensation reactions [9, 26, 28]. For most lignocellulosic biomass feedstocks, a good balance between high liquefaction efficiency and effective control of detrimental recondensation reactions can be obtained with approximately 3 % sulfuric acid loading.

Although extensive studies have been focused on acid-catalyzed liquefaction processes, there are a few reports on base-catalyzed liquefaction processes. Generally, base-catalyzed liquefaction requires a higher liquefaction temperature (around 250 °C) to achieve liquefaction efficiency comparable to those obtained from acid-catalyzed liquefaction [39, 40]. However, base-catalyzed liquefaction processes are less corrosive to metal equipment used in the liquefaction process and have lower acid numbers.

3.3.2.4 Effect of Temperature and Time

Temperature used in the liquefaction process is dependent on the type of catalysts, i.e., acid or base catalyst. Liquefaction temperature around 250 °C is needed for base-catalyzed liquefaction, while temperatures between 130 and 170 °C are needed for acid-catalyzed liquefaction to obtain comparable liquefaction efficiency. For an acid-catalyzed liquefaction process using binary mixtures of PEG 400/glycerol as liquefaction solvents, significant improvement of biomass liquefaction efficiency was observed as temperatures increased from 130 to 150 °C, beyond which little improvement was observed [6, 9, 26, 34, 35, 38, 41]. A rapid liquefaction of lignocellulosic biomass usually occurred in the first 15–30 min, after which the liquefaction proceeded at a much lower rate [6, 26, 34]. As discussed previously, this initial rapid liquefaction stage was largely due to the degradation of more accessible biomass components such as lignin, hemicellulose, and amorphous cellulose, while the later slow stage of the process mainly featured degradation of well-packed and less solvent-accessible crystalline cellulose. During the biomass liquefaction process, recondensation reactions among derivatives from biomass and/or liquefaction solvents always accompany and compete against the liquefaction reactions. These recondensation reactions become severe at high liquefaction temperatures and prolonged liquefaction time and significantly decrease biomass liquefaction efficiency. Overall, most lignocellulosic biomass feedstocks liquefied with acid catalysts at 150 °C for 90 min had liquefaction efficiencies higher than 90 % without significant occurrence of detrimental recondensation reactions. Comparable high liquefaction efficiencies have been obtained at 240–250 °C for 60 min in base-catalyzed biomass liquefaction processes [30].

3.3.3 Properties of Liquefaction-Derived Polyols

The polyols produced via liquefaction of lignocellulosic biomass are rich in hydroxyl groups and can be used directly to prepare PUs such as foams, adhesives, and films. Similar to oxypropylation-derived polyols, liquefaction-derived polyols produced from lignocellulosic biomass are typically characterized in terms of hydroxyl number and viscosity, all of which vary dynamically during the liquefaction process. However, while acid number is not a significant factor for oxpropylation-derived polyols, it needs to be considered for liquefaction-derived polyols. In general, during biomass liquefaction using petroleum-based polyhydric alcohols as solvents, the hydroxyl numbers and viscosities of polyols decrease, while the acid numbers of polyols increase. The decrease of polyol hydroxyl numbers can be ascribed to the consumption of hydroxyl groups by reactions such as oxidation and dehydration, while the increase of acid numbers can be explained by the formation of organic acids from the decomposition of lignocellulosic biomass and/or the oxidation of polyhydric solvents [34, 42, 43]. However, when crude glycerol was used as a liquefaction solvent, decreases in the acid number and increases in the viscosity of polyols were observed as the liquefaction temperature increased from 120 to 240 °C [22]. This contradictory phenomenon most likely resulted from esterification and transesterification reactions among crude glycerol components such as glycerol and fatty acids/fatty acid methyl esters during the liquefaction process [22, 44]. When liquefaction was conducted at 240 °C for prolonged reaction times, biomass liquefaction became dominant and the impact of liquefaction time on polyol properties (i.e., hydroxyl and acid numbers, viscosity)

Table 3.4 General properties of liquefaction-derived polyols from lignocellulosic biomass[a]

Lignocellulosic biomass	Polyol properties				References
	Liquefaction efficiency (%)	Acid number (mg KOH/g)	Hydroxyl number (mg KOH/g)	Viscosity (Pa.s)	
Enzymatic hydrolysis lignin	98	N/A[b]	249	N/A[b]	[50]
Cellulose or waste paper	55–99	19–30	360–396	2.6–3.9	[25, 36]
Agricultural crop residues[c]	60–95	15–30	109–430	1.0–1.7	[6, 7, 9, 26, 33, 35, 38]
Biorefinery residues[d]	84–98	28–34	137–586	0.4–3.0	[1, 9, 34, 41, 51–53]
Wood[e]	80–98	12–38	200–435	0.3–31.6	[32, 46, 47, 49]
Birch wood[f]	>99	24–41	112–204	N/A[b]	[39, 40]
Soybean straw[g]	65–75	<5	440–540	16–45	[22]

Note [a]Properties reported are for polyols produced from acid-catalyzed liquefaction using petroleum-derived polyhydric alcohols as solvents unless otherwise stated
[b]Not available
[c]Residues including wheat straw, rice straw, corn stover, cotton stalks, corn cobs, etc.
[d]Residues including bagasse, dried distillers grains (*DDG*), corn bran, date seeds, rapeseed cake residue, apple pomace, olive stone, etc.
[e]Wood including both softwood and hardwood species
[f]Base-catalyzed liquefaction using petroleum-derived polyhydric alcohols as liquefaction solvent
[g]Base-catalyzed liquefaction using biodiesel-derived crude glycerol as liquefaction solvent

were similar to those shown in conventional polyhydric alcohol-based liquefaction processes [22]. It was observed that the molecular weight of polyols usually increased at the early stages of the liquefaction process [27, 35]. This was explained by the release of a large number of macromolecules from biomass. With the gradual degradation of these macromolecules, the molecular weight of polyols decreased accordingly but it could increase again if significant recondensation reactions occurred at the late stages of the liquefaction process [27, 35].

Table 3.4 summarizes the general properties of polyols obtained from the liquefaction of lignocellulosic biomass, including wood, crop residues, and byproducts of biorefineries. With different types of biomass and liquefaction parameters, polyols show hydroxyl numbers ranging from approximately 100 to 600 mg KOH/g, acid numbers less than 40 mg KOH/g, and viscosities from 0.3 to 45 Pa.s. Studies suggest that liquefaction-derived polyols from lignocellulosic biomass are usually suitable for the production of rigid or semi-rigid PU foams but they can also be used for the preparation of PU adhesives [45, 46] and films [47–49].

3.3.4 Polyurethanes Produced from Polyols Derived from Liquefaction

Polyols derived from liquefaction of lignocellulosic biomass have been tested for the production of rigid or semi-rigid PU foams. Acid-catalyzed liquefaction-derived polyols usually have high acid numbers of up to 40 mg KOH/g, which are not favorable for the production of PU foams, as the residual acids of polyols reduce the catalytic activity of tertiary amines by the formation of ammonium salts during PU foaming [54]. In order to produce PU foams with appropriate properties from polyols derived by acid-catalyzed liquefaction, neutralization of polyols with a base such as NaOH or MgO is necessary [6, 52, 55]. The neutralized polyols can be used as partial substitutes for petroleum-derived polyols or alone to produce PU foams. Studies showed that PU foams produced from mixtures of biomass liquefaction- and petroleum-derived polyols had improved compressive strength and thermal stability over those from petroleum-derived polyols alone [6, 52, 56]. These property improvements can be largely attributed to the increased hard segment contents and crosslinking densities in PU foams produced from biomass liquefaction-derived polyols. For PU foams produced from 100 % biomass liquefaction-derived polyols, their properties depend on both polyol properties and foaming formulations. For example, PU foams from polyols obtained via combined liquefaction of wood and starch at different ratios, showed densities of around of 30 kg/m^3, compressive strengths of 80–150 kPa, and elastic moduli of 3–10 MPa [55]. PU foams from polyols produced by liquefaction of soybean straw in crude glycerol had densities ranging from 33 to 37 kg/m^3 and compressive strengths from 148 to 227 kPa [22]. Foaming formulations, such as the NCO/OH molar ratio and catalyst loading, also have important impacts on the properties of biomass liquefaction-derived PU foams [8, 35]. High NCO/OH molar ratios helped to produce PU foams with high mechanical strength and high thermal stability, as a result of the increased hard segment content and crosslinking density in polymer networks [35]. However, excessive amounts of isocyanates in foaming formulations could be detrimental to foam properties due to the incomplete curing of isocyanates [35].

Flexible and highly resilient PU foams have also been produced from polyols obtained via lignocellulosic biomass liquefaction, in which petroleum-derived polyether polyols with high molecular weights were used as liquefaction solvents. In a study on the liquefaction of tannin-containing bark and corn starch for PU foam production, PPG 4000 (MW: 4000 g/mol) provided the optimal balance between high liquefaction efficiency and maximum foam resilience [57]. The study also indicated that the bark tannin and starch contents in the liquefaction system also substantially affected compressive strengths, densities, and resilience of the resulting PU foams. The higher the tannin content in the bark, the higher the density and resilience values of PU foams, and the higher the starch content in the feedstock, the higher the densities and compressive strengths while the lower the resilience values [57].

Besides applications for PU foams, biomass liquefaction-derived polyols also have been used to produce other forms of PUs, such as adhesives [45, 46], resins [58], and films [47–49]. Similar to PU foams, the formulations, such as the type of isocyanates and the molar ratio of NCO/OH, have also been shown to affect properties of PU adhesives, resins, and films, including the mechanical and thermal stability properties, the gel time and wood bonding strength of adhesives, and the crosslink density and sol fraction (i.e., soluble fraction) of films. In a study of synthesis of PU resins based on liquefied benzylated wood, thermal stability and surface properties of PUs were dependent on the isocyanate type [i.e., toluene diisocyanate (TDI), isophorone diisocyanate (IPDI), and hexamethylene diisocyanate (HDI)] [58]. PU adhesives derived from liquefied Taiwan acacia or China fir wood using Desmodur L (adduct of toluene diisocyanate with trimethylol propane) had better dry and wet bonding strength than those using polymeric methylene diphenylene diisocyanate (pMDI) or Desmodur N (trimer of hexamethylene diisocyanate) [46]. In addition, PU adhesives from liquefied Taiwan acacia showed better bonding strength than those from liquefied China fir. With a higher NCO/OH ratio in the PU formulation, increased bonding strength of PU adhesives could be obtained. In studies of the mechanical properties and network structures of PU films prepared from polyols derived from the liquefaction of woody biomass, it was discovered that liquefied wood in polyols acted as crosslinkers in PU networks, leading to increased tensile strength, glass transition temperature, and thermal stability of PU films [48, 49]. The crosslink density of PU films increased from 110 to 1360 mol/m^3 as the molar ratio of NCO/OH increased from 0.6 to 1.4. In contrast, the weight percentage of the sol fraction of PU films decreased from 39.2 to 3.1 % with an increase of the molar ratio of NCO/OH from 0.6 to 1.4 [48].

3.4 Conclusion and Remarks

As the most abundantly available biomass, lignocellulosic biomass has great potential to produce polyols via either oxypropylation or atmospheric liquefaction processes. Oxypropylation is usually conducted under high pressure and temperature using KOH as a catalyst in a process that uses propylene oxide to introduce poly(propylene oxide) to biomass structures. Liquefaction processes are commonly conducted at elevated temperatures (ca. 150–250 °C) under atmospheric pressure and can be either acid- or alkali-catalyzed, with the former being more common. The reaction parameters in both processes have significant effects on the properties of the derived polyols, which can further affect performance of the resulting PUs. Polyols derived from oxypropylation or liquefaction of lignocellulosic biomass have been used in laboratory tests to produce PU foams with performance comparable to analogs from petroleum-based polyols. In some cases, the biomass liquefaction-derived PU foams have exhibited similar mechanical properties but better thermal stability compared to some of their petroleum-derived analogs. However, lignocellulosic biomass derived polyols face some challenges for wide applications for other PU products, such as flexible foams, adhesives, and coatings. A recent study of the combined use of liquefaction and oxypropylation reported that polyols with higher molecular weights and lower viscosities were produced via the liquefaction of biomass followed by oxypropylation [59]. The produced polyols were used for the production of rigid PU foams, which showed physical properties as good as or better than those from petroleum-based polyols. Thus, a combination of liquefaction and oxypropylation of lignocellulosic biomass may be promising for producing polyols for the development of high quality PU flexible foams, adhesives, and coatings.

Additionally, large quantities of petroleum-derived solvents are used in the biomass liquefaction process. This inevitably reduces the bio-content and renewability of biomass liquefaction-derived polyols and PUs. The use of bio-based liquefaction solvents is expected to play an important role in resolving or alleviating these issues. Despite these challenges, the inevitable depletion of the world's petroleum resources demands continued efforts to advance the technologies and economics of lignocellulosic biomass-based polyols and PUs.

References

1. Briones R, Serrano L, Labidi J (2012) Valorization of some lignocellulosic agro-industrial residues to obtain biopolyols. J Chem Technol Biot 87:244–249
2. Cateto CA, Barreiro MF, Rodrigues AE, Belgacem MN (2009) Optimization study of lignin oxypropylation in view of the preparation of polyurethane rigid foams. Ind Eng Chem Res 48:2583–2589
3. Yoshida H, Mörck R, Kringstad KP, Hatakeyama H (1987) Kraft lignin in polyurethanes I. Mechanical properties of polyurethanes from a kraft lignin-polyether triol-polymeric MDI system. J Appl Polym Sci 34:1187–1198
4. Yoshida H, Mörck R, Kringstad KP, Hatakeyama H (1990) Kraft lignin in polyurethanes. II. Effects of the molecular weight of kraft lignin on the properties of polyurethanes from a kraft lignin-polyether triol-polymeric MDI system. J Appl Polym Sci 40:1819–1832
5. Cateto CA, Barreiro MF, Rodrigues AE (2008) Monitoring of lignin-based polyurethane synthesis by FTIR-ATR. Ind Crop Prod 27:168–174
6. Chen F, Lu Z (2009) Liquefaction of wheat straw and preparation of rigid polyurethane foam from the liquefaction products. J Appl Polym Sci 111:508–516
7. Wang T, Li D, Wang L, Yin J, Chen XD, Mao Z (2008) Effects of CS/EC ratio on structure and properties of polyurethane foams prepared from untreated liquefied corn stover with PAPI. Chem Eng Res Des 86:416–421
8. Gao L-L, Liu Y-H, Lei H, Peng H, Ruan R (2010) Preparation of semirigid polyurethane foam with liquefied bamboo residues. J Appl Polym Sci 116:1694–1699
9. Hassan EM, Shukry N (2008) Polyhydric alcohol liquefaction of some lignocellulosic agricultural residues. Ind Crop Prod 27:33–38
10. Pavier C, Gandini A (2000) Oxypropylation of sugar beet pulp. 1. Optimisation of the reaction. Ind Crop Prod 12:1–8
11. Xue BL, Wen JL, Sun RC (2014) Lignin-based rigid polyurethane foam reinforced with pulp fiber: synthesis and characterization. ACS Sustain Chem Eng 2:1474–1480
12. Gandini A (2011) The irruption of polymers from renewable resources on the scene of macromolecular science and technology. Green Chem 13:1061–1083

13. Chung H, Washburn NR (2012) Chemistry of lignin-based materials. Green Mater 1:137–160
14. Nadji H, Bruzzèse C, Belgacem MN, Benaboura A, Gandini A (2005) Oxypropylation of lignins and preparation of rigid polyurethane foams from the ensuing polyols. Macromol Mater Eng 290:1009–1016
15. Gandini A, Belgacem MN (2008) Partial or total oxypropylation of natural polymers and the use of the ensuing materials as composites or polyol macromonomers. In: Belgacem MN, Gandini A (eds) Monomers, polymers and composites from renewable resources, 1st edn. Elsevier, Oxford
16. Evtiouguina M, Barros-Timmons A, Cruz-Pinto JJ, Neto CP, Belgacem MN, Gandini A (2002) Oxypropylation of cork and the use of the ensuing polyols in polyurethane formulations. Biomacromolecules 3:57–62
17. Aniceto JP, Portugal I, Silva CM (2012) Biomass-based polyols through oxypropylation reaction. ChemSusChem 5:1358–1368
18. Pavier C, Gandini A (2000) Oxypropylation of sugar beet pulp. 2. Separation of the grafted pulp from the propylene oxide homopolymer. Carbohyd Polym 42:13–17
19. Gandini A, Belgacem MN (2002) Recent contributions to the preparation of polymers derived from renewable resources. J Polym Environ 10:105–114
20. Matos M, Barreiro MF, Gandini A (2010) Olive stone as a renewable source of biopolyols. Ind Crop Prod 32:7–12
21. Serrano L, Alriols MG, Briones R, Mondragón I, Labidi J (2010) Oxypropylation of rapeseed cake residue generated in the biodiesel production process. Ind Eng Chem Res 49:1526–1529
22. Hu S, Wan C, Li Y (2012) Production and characterization of biopolyols and polyurethane foams from crude glycerol based liquefaction of soybean straw. Bioresour Technol 103:227–233
23. De Menezes AJ, Pasquini D, Curvelo AAS, Gandini A (2007) Novel thermoplastic materials based on the outer-shell oxypropylation of corn starch granules. Biomacromolecules 8:2047–2050
24. Yao Y, Yoshioka M, Shiraishi N (1993) Combined liquefaction of wood and starch in a polyethylene glycol/glycin blended solvent. Mokuzai Gakkaishi 39:930–938
25. Lee S-H, Teramoto Y, Shiraishi N (2002) Biodegradable polyurethane foam from liquefied waste paper and its thermal stability, biodegradability, and genotoxicity. J Appl Polym Sci 83:1482–1489
26. Zhang H, Ding F, Luo C, Xiong L, Chen X (2012) Liquefaction and characterization of acid hydrolysis residue of corncob in polyhydric alcohols. Ind Crop Prod 39:47–51
27. Kobayashi M, Asano T, Kajiyama M, Tomita B (2004) Analysis on residue formation during wood liquefaction with polyhydric alcohol. J Wood Sci 50:407–414
28. Zhang H, Pang H, Shi J, Fu T, Liao B (2012) Investigation of liquefied wood residues based on cellulose, hemicellulose, and lignin. J Appl Polym Sci 123:850–856
29. Zhang T, Zhou Y, Liu D, Petrus L (2007) Qualitative analysis of products formed during the acid catalyzed liquefaction of bagasse in ethylene glycol. Bioresour Technol 98:1454–1459
30. Hu S, Luo X, Li Y (2014) Polyols and polyurethanes from the liquefaction of lignocellulosic biomass. ChemSusChem 7:66–72
31. Xu J, Jiang J, Hse C, Shupe TF (2012) Renewable chemical feedstocks from integrated liquefaction processing of lignocellulosic materials using microwave energy. Green Chem 14:2821–2830
32. Kurimoto Y, Doi S, Tamura Y (1999) Species effect on wood liquefaction in polyhydric alcohols. Holzforschung 53:617–622
33. Liang L, Mao Z, Li Y, Wan C, Wang T, Zhang L, Zhang L (2006) Liquefaction of crop residues for polyol production. BioResources 1:248–256
34. Lee S-H, Yoshioka M, Shiraishi N (2000) Liquefaction of corn bran (CB) in the presence of alcohols and preparation of polyurethane foam from its liquefied polyol. J Appl Polym Sci 78:319–325

35. Yan Y, Pang H, Yang X, Zhang R, Liao B (2008) Preparation and characterization of water-blown polyurethane foams from liquefied cornstalk polyol. J Appl Polym Sci 110:1099–1111
36. Yamada T, Ono H (1999) Rapid liquefaction of lignocellulosic waste by using ethylene carbonate. Bioresour Technol 70:61–67
37. Hu S, Luo X, Wan C, Li Y (2012) Characterization of crude glycerol from biodiesel plants. J Agr Food Chem 60:5915–5921
38. Wang H, Chen H (2007) A novel method of utilizing the biomass resource: rapid liquefaction of wheat straw and preparation of biodegradable polyurethane foam (PUF). J Chin Inst Chem Eng, 38:95–102
39. Maldas D, Shiraishi N (1996) Liquefaction of wood in the presence of polyol using NaOH as a catalyst and its application to polyurethane foams. Int J Polym Mater 33:61–71
40. Alma MH, Shiraishi N (1998) Preparation of polyurethane-like foams from NaOH-catalyzed liquefied wood. Holz Roh Werkst 56:245–246
41. Briones R, Serrano L, Llano-Ponte R, Labidi J (2011) Polyols obtained from solvolysis liquefaction of biodiesel production solid residues. Chem Eng J 175:169–175
42. Yamada T, Ono H (2001) Characterization of the products resulting from ethylene glycol liquefaction of cellulose. J Wood Sci 47:458–464
43. Yamada T, Aratani M, Kubo S, Ono H (2007) Chemical analysis of the product in acid-catalyzed solvolysis of cellulose using polyethylene glycol and ethylene carbonate. J Wood Sci 53:487–493
44. Luo X, Hu S, Zhang X, Li Y (2013) Thermochemical conversion of crude glycerol to biopolyols for the production of polyurethane foams. Bioresour Technol 139:323–329
45. Juhaida MF, Paridah MT, Hilmi MM, Sarani Z, Jalaluddin H, Zaki ARM (2010) Liquefaction of kenaf (*Hibiscus cannabinus* L.) core for wood laminating adhesive. Bioresour Technol 101:1355–1360
46. Lee W-J, Lin M-S (2008) Preparation and application of polyurethane adhesives made from polyhydric alcohol liquefied Taiwan acacia and China fir. J Appl Polym Sci 109:23–31
47. Kurimoto Y, Koizumi A, Doi S, Tamura Y, Ono H (2001) Wood species effects on the characteristics of liquefied wood and the properties of polyurethane films prepared from the liquefied wood. Biomass Bioenergy 21:381–390
48. Kurimoto Y, Takeda M, Doi S, Tamura Y, Ono H (2001) Network structures and thermal properties of polyurethane films prepared from liquefied wood. Bioresour Technol 77:33–40
49. Kurimoto Y, Takeda M, Koizumi A, Yamauchi S, Doi S, Tamura Y (2000) Mechanical properties of polyurethane films prepared from liquefied wood with polymeric MDI. Bioresour Technol 74:151–157
50. Jin Y, Ruan X, Cheng X, Lü Q (2011) Liquefaction of lignin by polyethyleneglycol and glycerol. Bioresour Technol 102:3581–3583
51. Yu F, Le Z, Chen P, Liu Y, Lin X, Ruan R (2008) Atmospheric pressure liquefaction of dried distillers grains (DDG) and making polyurethane foams from liquefied DDG. Appl Biochem Biotechnol 148:235–243
52. Liu J, Chen F, Qiu M (2009) Liquefaction of bagasse and preparation of rigid polyurethane foam from liquefaction products. J Biobased Mater Bio 3:401–407
53. Briones R, Serrano L, Younes RB, Mondragon I, Labidi J (2011) Polyol production by chemical modification of date seeds. Ind Crop Prod 34:1035–1040
54. Ionescu M (2005) Chemistry and technology of polyols for polyurethanes. iSmithers Rapra Publishing, UK
55. Yao Y, Yoshioka M, Shiraishi N (1995) Rigid polyurethane foams from combined liquefaction mixtures of wood and starch. Mokuzai Gakkaishi 41:659–668
56. Hakim AAA, Nassar M, Eman A, Sultan M (2011) Preparation and characterization of rigid polyurethane foam prepared from sugar-cane bagasse polyol. Mater Chem Phys 129:301–307

57. Ge J, Zhong W, Guo Z, Li W, Sakai K (2000) Biodegradable polyurethane materials from bark and starch. I. Highly resilient foams. J Appl Polym Sci 77:2575–2580
58. Wei Y, Cheng F, Li H, Yu J (2004) Synthesis and properties of polyurethane resins based on liquefied wood. J Appl Polym Sci 92:351–356
59. Yoshioka M, Nishio Y, Saito D, Ohashi H, Hashimoto M, Shiraishi N (2013) Synthesis of biopolyols by mild oxypropylation of liquefied starch and its application to polyurethane rigid foams. J Appl Polym Sci 130:622–630

Chapter 4
Polyols and Polyurethanes from Protein-Based Feedstocks

Abstract Feedstocks that have high protein contents, such as soy protein, are promising materials for extensive polyol and polyurethane (PU) applications, such as foams, films, and coatings, due to the characteristic structures and properties of proteins. Currently, most research has been focused on the direct use of these protein-based feedstocks in combination with polymers for PU production. Although proteins have multiple reactive functional groups, such as amino and carboxyl groups, reports on the modification of protein-based feedstocks for the production of liquid polyols are limited. This chapter reviews sources, compositions, structures, and processing of protein-based feedstocks; synthetic methods and properties of protein-based polyols; and performance and applications of the derived PUs.

Keywords Polyurethanes · Soy meal · Soy protein isolate · Soy protein concentrate · Defatted soy flour · Soy dreg · Distillers dried grains · Wheat gluten · Zein · Algal biomass

4.1 Introduction

Proteins are polypeptides with long and complex structures. Polypeptide is formed by linking many α-amino acids via amide bonds (i.e., peptide bonds). Figure 4.1 shows the structure of amide bonds of a polypeptide. Proteins have four levels of structural complexity: primary, secondary, tertiary, and quaternary. The amino acid sequence of polypeptide chains of a protein molecule makes up the primary structure of the protein. Secondary protein structures are formed mainly via hydrogen bonding of amino acids, which causes folding or coiling within the protein; whereas, tertiary structures are formed via interactions of side-chain groups of amino acids, such as disulfide linkages, ionic interactions, hydrogen bonding, and hydrophobic interaction. Quaternary structures involve the spatial arrangement of two or more polypeptide chains for the formation of a larger protein.

© The Author(s) 2015
Y. Li et al., *Bio-based Polyols and Polyurethanes*,
SpringerBriefs in Green Chemistry for Sustainability,
DOI 10.1007/978-3-319-21539-6_4

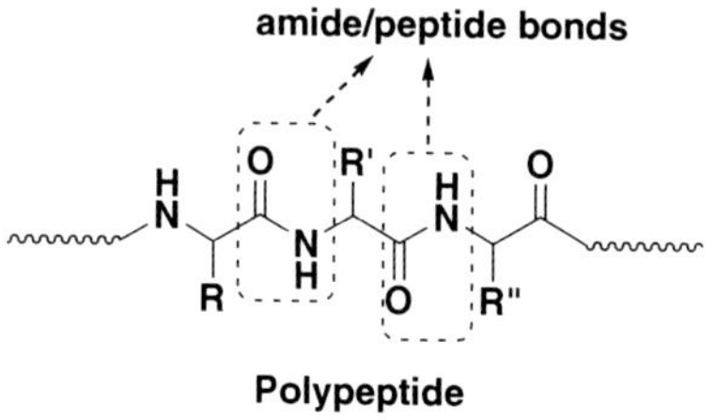

Fig. 4.1 Structure of amide bonds of a polypeptide

Proteins contain many functional groups, such as amino ($-NH_2$), carboxyl ($-COOH$) and hydroxyl ($-OH$) groups, and disulfide bonds ($-S-S-$) [1], which can react with other functional groups to generate polymeric materials with new properties. Physical (e.g., heat, mild alkali, and pressure treatments), chemical (e.g., acetylation, succinylation, and deamidation), and enzymatic approaches are commonly used to modify proteins [2, 3]. Denaturation is a modification process in which only the secondary, tertiary, or quaternary structures of a protein are changed while its amino acid sequence remains unchanged [3]. During denaturation, the protein structure becomes unfolded via the disruption of inter- and intra-molecular interactions such as hydrogen and disulfide bonds.

Due to their diverse structures, proteins exhibit a variety of physicochemical properties and biological functions. Solubility is one of the most important properties of proteins. It can be affected by the protein's hydrophobic, hydrophilic, and steric properties; its size and electrical charge; pH of the aqueous solution; and the type and ionic strength of various salts [2]. Soy protein has a minimum solubility in aqueous solution at its isoelectric region (i.e., pH = 4.2–4.6). When the pH is above or below the isoelectric region, the solubility of soy protein increases dramatically. Thus, adjusting pH to the isoelectric region is a commonly used method for the production of soy protein. In salt solutions, such as NaI, NH_4NO_3, NH_4Br, NaCl, $(NH_4)_2SO_4$, and Na_2SO_4, sulfate salts were found to have the strongest influence in reducing the solubility of soy protein [2]. Proteins have low solubility in organic solvents because of their complex macromolecular structures and strong intermolecular and intramolecular interactions. This can hinder functionalization of proteins for industrial uses.

Feedstocks that have high protein contents are abundant in nature and can be obtained from a variety of sources. They include: (1) soy protein products, such as soy meal, soy protein isolate (SPI), soy protein concentrate (SPC), defatted soy flour, and soy dreg; (2) corn protein products, such as corn gluten meal, corn gluten feed, and distillers dried grains (DDG) or distillers dried grains with solubles (DDGS); (3) wheat gluten; and (4) algal biomass. The components and characteristics of these protein-based feedstocks vary as discussed in Sect. 4.2.

4.2   Protein-Based Feedstocks

4.2.1   Soy Protein Products

Soy protein products have different protein contents depending on the preparation process. Table 4.1 shows typical chemical compositions of five soy protein products.

Defatted soy meal is a byproduct of the soybean oil extraction process in which hexane is usually used. Hexane can be removed from the insoluble fraction via a desolventizing process, such as a vapor or flash desolventizing process [2]. Figure 4.2 shows a schematic diagram for the production of defatted soy meal.

Soy flour is finely ground soybeans or soy meal and has a particle size smaller than 100 mesh. Depending on the oil content, soy flour can be classified into four categories: full-fat, high-fat, low-fat, and defatted. Full-fat soy flour is obtained from soybeans without oil extraction and contains about 18–20 % oil. Defatted soy flour is obtained from defatted soy meal and usually contains less than 1 % oil (Table 4.1). Low- and high-fat soy flours are obtained by adding soybean oil to defatted soy flour, and contain about 4.5–9.0 and 10 % or more oil, respectively [8].

As shown in Fig. 4.3, soy protein concentrates (SPCs) are produced from defatted soy meal by three common processing methods: (1) acid treatment; (2) aqueous alcohol leaching; and (3) moist heat and water leaching [2]. During acid treatment of soy meal, the pH of the soy meal water solution is first adjusted to about 4.5 to precipitate soy proteins, and insoluble soy proteins are then collected by centrifugation. Finally, SPCs are obtained as powders by neutralization and

Table 4.1 Typical chemical composition of five soy protein products

Composition (wt%)	Soy meal [4]	Defatted soy flour [5]	Soy protein concentrates [6]	Soy protein isolates [6]	Soy dreg [7]
Protein	50.3	53.3	72.0	96.0	12.0
Fat	0.5	0.3	1.0	0.1	–
Crude fiber	8.7	3.3	4.5	0.1	–
Ash	4.4	7.8	5.0	3.5	2.0
Carbohydrate	36.1	35.3	17.5	0.3	86.0

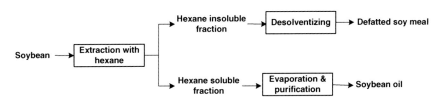

Fig. 4.2 A schematic diagram for hexane extraction of soybean oil and production of defatted soy meal

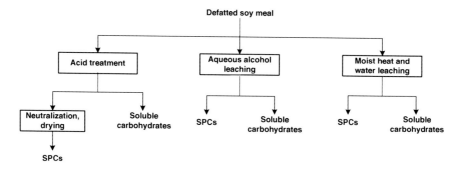

Fig. 4.3 A schematic diagram for the production of soy protein concentrates

drying of the collected soy proteins. During the aqueous alcohol leaching process, SPCs consisting of alcohol-insoluble soy proteins and polysaccharides are obtained by first removing alcohol-soluble carbohydrates, and then desolventizing and drying. During the moist heat and water leaching process, soy proteins denature and become water-insoluble after defatted soy meal is heated in the presence of moisture. SPCs are then obtained through water leaching to remove water-soluble carbohydrates followed by drying.

Traditionally, soy protein isolates (SPIs) are produced by mild alkali extraction of defatted soy meal followed by precipitation at the isoelectric region of soy proteins (Fig. 4.4). Defatted soy meal is first dispersed in a mild alkali solution (e.g.,

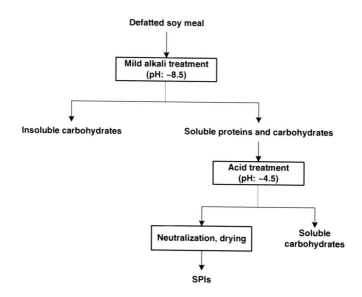

Fig. 4.4 A schematic diagram for the production of soy protein isolates

sodium hydroxide solution) with a pH of around 8.5. The insoluble carbohydrates are then removed by centrifugation and the supernatant, containing soluble proteins carbohydrates, is collected. Subsequently, after adjusting the pH of the supernatant to about 4.5, soy proteins are precipitated and separated. Powder SPIs are obtained after neutralization, washing, and drying. Additionally, SPIs in an isoelectric form are also commonly obtained via washing and drying of soy protein precipitates without neutralization. As both soluble and insoluble carbohydrates are removed during the production of SPIs, the protein content of SPIs is approximately 90 % or higher, which is generally higher than that of SPCs (Table 4.1).

Soy dreg is an abundantly available byproduct from the soy protein industry. Its price is much cheaper than that of SPIs. Soy dreg mainly contains protein, cellulose, and other carbohydrates, in which the content of soy protein is about 12 % (Table 4.1).

4.2.2 Corn Protein Products

Corn protein products mainly include corn gluten meal, corn gluten feed, distillers dried grains (DDG) and distillers dried grains with solubles (DDGS), which are obtained from corn wet-milling and dry-milling processes, respectively [9]. Figure 4.5 shows a schematic diagram for the production of these corn protein products. Usually, corn gluten meal contains approximately 65 % protein, corn gluten feed contains about 23 % protein, and DDGS contains about 27 % protein (Table 4.2). Zein, a functional prolamine protein found in corn, can be extracted from the above corn protein products and has been used to formulate PUs. Commercial zein is commonly produced from corn gluten meal [10]. Zein is insoluble in water while soluble in the presence of alcohol or a high concentration of urea or alkali (pH $\geq$ 11) [9].

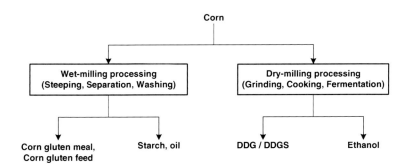

Fig. 4.5 A schematic diagram for the production of corn protein products

Table 4.2 Chemical composition of corn and wheat protein products

Composition (wt%)	Corn gluten meal [9]	Corn gluten feed [9]	Distillers dried grains with solubles (DDGS) [9]	Wheat gluten [11]
Protein	65.0	23.0	27.0	83.5
Starch	20.0	27.0	–	6.2
Oil	4.0	2.4	13.0	N/A[c]
Ash	1.0	1.0	4.0	N/A[c]
Others	10.0[a]	46.6[a]	56.0[b]	N/A[c]

Note [a]Others include fiber, pentosan, phytic acid, and soluble sugar
[b]In addition to the others specified in a, it contains glycerol, organic acids, and other products in the dry-milling ethanol process
[c]Not available

4.2.3 Wheat Gluten

Wheat gluten is obtained by washing wheat dough with water. During the washing process, most starch and water-soluble components are removed. Wheat gluten generally contains about 75–85 % protein, 5–10 % lipids, and 5–20 % carbohydrates and others [12] (Table 4.2). Wheat gluten has two major types of proteins: glutenin and gliadin, in which, glutenin provides elastic properties while gliadin primarily provides viscous properties [13]. The interaction of these two proteins endows wheat protein with viscoelastic properties when wheat gluten is mixed with water. Wheat gluten has been successfully used for the production of bioplastics with promising performance [14].

4.2.4 Algal Biomass

Algal biomass can be classified into two categories: macroalgal and microalgal biomass. Both of them are mainly composed of proteins, lipids, and carbohydrates. The chemical composition of macroalgae and microalgae varies among different species (Table 4.3), and can be affected by the culture medium and environmental conditions. In recent decades, much attention has been given to microalgae as a promising feedstock for the production of biodiesel due to its rapid growth rate and high oil content [15]. Raceway ponds or tubular photobioreactors are commonly used for large-scale production of microalgae. After harvesting the algal biomass by centrifugation, filtration, or gravity sedimentation [16], microalgae oil can be extracted for the production of biodiesel. Value-added utilization of the remaining biomass residue with high protein content, including production of methane by anaerobic digestion [15] and composites [17] (Fig. 4.6), could help improve the economics of algae production.

Table 4.3 Chemical composition of different algal biomass

Species	Macroalgal biomass [18]			Microalgal biomass [19]	Microalgal biomass residue [20]
	Derbesiaa	Cladophora[a]	Cladophora[b]	Chlorella sp.	N. salina
Protein (wt%)	23.1	11.7	28.4	9.3	47.9
Lipid (wt%)	11.1	3.5	5.6	59.9	14.6
Ash (wt%)	37.1	38.6	18.9	4.9	N/A[c]
Carbohydrates (wt%)	28.7	46.2	47.1	25.9	4.1

Note [a]Macroalgae was cultured in marine water
[b]Macroalgae was cultured in fresh water
[c]Not available

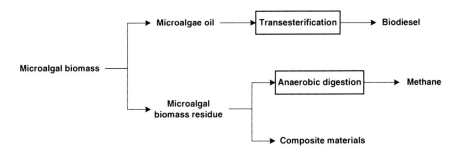

Fig. 4.6 A schematic diagram for the production of microalgal biomass products

4.3 Polyols from Protein-Based Feedstocks

Two methods have typically been used to produce polyols from protein feedstocks: (1) modification of proteins into polyols; and (2) liquefaction of proteins. The production of polyols has been reported via the modification of proteins with polyethylene glycol (PEG), such as the addition reaction of PEG-aldehyde with protein and the substitution reaction of cyanuric chloride-modified PEG with protein, with the latter being more widely used [21] (Scheme 4.1). Soy meal-based polyols produced through oxypropylation of hydrolyzate of soy meal had a hydroxyl number of 442 mg KOH/g and an amine value of 102 mg KOH/g [22], while algal protein-based polyols produced through reactions of algal hydrolyzate with ethylene diamine and then with ethylene carbonate (Scheme 4.2) showed a hydroxyl number of 422 mg KOH/g, an acid number of 3 mg KOH/g, and an amine value of 26 mg KOH/g [23].

Scheme 4.1 Synthesis of protein-based polyols by the modification of PEG with the use of cyanuric chloride [21]

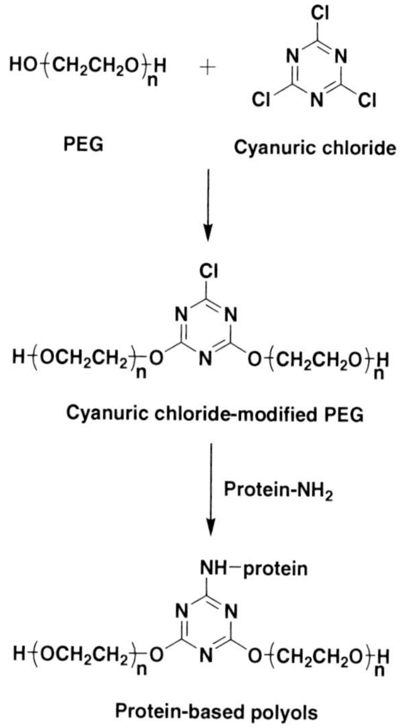

HO$\{$CH$_2$CH$_2$O$\}_n$H + Cyanuric chloride

PEG   **Cyanuric chloride**

H$\{$OCH$_2$CH$_2$$\}_n$O—[triazine ring]—O$\{CH_2CH_2O\}_n$H

Cyanuric chloride-modified PEG

Protein-NH$_2$

NH—protein

H$\{$OCH$_2$CH$_2$$\}_n$O—[triazine ring]—O$\{CH_2CH_2O\}_n$H

Protein-based polyols

Scheme 4.2 Production of algal protein-based urethane polyol using glycine (one of the amino acids in algal hydrolyzate) [23]

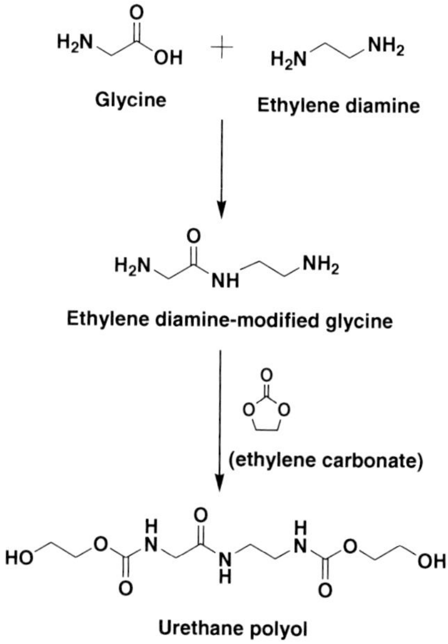

Glycine + **Ethylene diamine**

Ethylene diamine-modified glycine

(ethylene carbonate)

Urethane polyol

Liquefaction is a useful method for polyol synthesis from lignocellulosic biomass, and has been shown in one study to be effective in utilizing protein-based feedstocks for polyol production as well. Liquefaction of distillers dried grains (DDG) under atmospheric pressure in ethylene carbonate using sulfuric acid as a catalyst for the production of bio-based polyols has been reported by Yu et al. [24]. With an increase of liquefaction temperature (150–170 °C) and time (0.5–3 h), the sulfuric acid loading (1–5 wt%), and weight ratio of the liquefaction solvent (i.e., ethylene carbonate) to DDG (3:1–5:1), increased liquefaction yields were observed [24]. However, a high sulfuric acid loading may cause condensation of degraded residues and increase the viscosities of polyols. Additionally, a high weight ratio of liquefaction solvent to DDG will reduce the bio-based content of the produced polyols. The optimum liquefaction conditions of DDG for producing polyols were a temperature of 160 °C, time of 2 h, sulfuric acid loading of 3 %, and weight ratio of ethylene carbonate to DDG of 4:1, under which the liquefaction yield was 94.6 % and the resulting polyol had a hydroxyl number of 158 mg KOH/g [24].

4.4 Polyurethanes from Protein-Based Feedstocks

4.4.1 Foams

Currently, only a few studies have reported on the production of polyurethane (PU) foams from protein-based polyols [21–24]. For example, liquefied DDG-derived polyols were tested for the production of flexible and rigid PU foams [24]. However, performance characteristics, such as density, compressive strength, and indentation force deflection, of the produced PU foams, have not been reported. Recently, soy meal- and algal protein-based polyols were synthesized for PU foam applications [22, 23]. During the foam formulations, both soy meal- and algal protein-based polyols showed self-catalytic properties as a result of the presence of tertiary amine groups in their protein structures. This was confirmed by the fact of shorter cream time, gel time, rise time, and tack-free time for PU foams from soy meal- and algal protein-based polyols than those for the reference foams from commercial polyether polyols. PU foams prepared from soy meal-based polyols had properties comparable to those of the reference foam [22], while PU foams prepared from algal protein-based polyols showed relatively inferior properties compared to those of the reference foam [23]. The performance difference from these two protein-based PU foams resulted from the chemical structures of soy meal- and algal protein-based polyols which were synthesized using different approaches [22, 23].

Much attention has been paid to the direct use of protein-based feedstocks, especially soy protein products, to blend with polyether polyols for the production of rigid and flexible PU foams. The potential for using soy proteins to produce environmentally friendly PU foams with comparable or improved properties has been demonstrated. Blending of defatted soy flour and polyether polyol (e.g.,

Voranol 490, Dow Chemical) obtained rigid PU foams which showed density, compressive strength, and thermal conductivity in the range of 30.5–45.2 kg/m^3, 186–265 kPa, and 24.8–29.3 mW.m^{-1}.K^{-1}, respectively [25]. As the content of defatted soy flour increased from 0 to 30 % in the PU formulations, the density and compressive strength of foams increased. Compared to PU foams from polyether polyols, foams that incorporated defatted soy flour showed slightly increased thermal conductivities. Good dimensional stability with volume changes of less than 2 % was observed for all defatted soy flour-based PU foams [25]. Similar to defatted soy flour-based PU foams, rigid PU foams produced from a mixture of soy protein isolate (SPI) and polyether polyols had slightly higher thermal conductivities than that of their analogue without SPI [26]. SPI-based PU foams also showed good dimensional stability with volume changes of less than 2 % [26].

Soy proteins contain many active groups, including hydroxyl, amine, and carboxylic acid groups as well as disulfide bonds, which may be buried inside the complex structures of proteins [1]. To expose these groups and improve their reactivity with isocyanates in the foaming process, soy proteins have been modified. For example, soybean meal/SPI modified by an alkali treatment was used for the synthesis of rigid PU foams which had higher compressive strength, lower thermal conductivity, and better dimensional stability than rigid foams from unmodified soybean meal/SPI [4]. Table 4.4 shows properties of rigid PU foams from 30 % modified and unmodified soybean meal/SPI. Higher contents of functional groups in modified soybean meal and SPI may contribute to improved performance of the PU foams. Compared to SPI-based PU foams, soy meal-based PU foams showed better properties. This is most likely due to the higher carbohydrate content of soybean meal compared to SPI (Table 4.1) [4]. Fourier transform infrared spectroscopy (FT-IR) analysis indicated that soybean meal/SPI participated in the reaction with isocyanates during the synthesis of PU foams rather than just functioning as a filler [4].

Table 4.4 Properties of PU foams from 30 % modified and unmodified soybean meal/SPI [4]

	Density (kg/m^3)	Compressive strength (kPa)	Thermal conductivity (mW.m^{-1}.K^{-1})	Dimensional stability[a] (% volume change)		
				Day 1	Day 7	Day 14
Modified soybean meal	67.6	348	26.5	0.36	0.46	0.81
Unmodified soybean meal	63.8	242	31.2	0.78	0.98	1.21
Modified SPI	65.3	262	28.6	0.60	0.78	1.02
Unmodified SPI	63.7	221	32.1	1.15	1.18	1.22

Note [a]Condition: 75 °C and 5 % relative humidity

SPI-based flexible PU foams produced from blends of SPI and glycerol-propylene oxide polyether triol showed increased density (27–42 kg/m^3), improved resilience (22–34 %), higher indentation hardness (13.5–20.0 kPa at 65 % deflection), and higher comfort factor (2.0–3.3) as SPI content increased from 0 to 40 % [27]. The improved properties of SPI-based foams suggested significant potential for SPI in flexible foams for packaging and cushioning applications. Blending of soy protein concentrate (SPC) (≤20 %) or defatted soy flour (≤20 %) with glycerol-propylene oxide polyether triol also increased both the density and resilience value of the PU foams [28]. Compared to SPI-based flexible foam, SPC- and defatted soy flour-based foams showed lower density and resilience value with the same amount of protein feedstock in the formulation [28]. This was mainly attributed to the different composition of these three soy protein products (Table 4.1).

4.4.2 Blend Films/Coatings

Due to its good film-forming ability, biodegradability, biocompatibility, and processability [29], soy protein has been used for the production of edible and non-edible films and coatings. However, soy protein-based films are brittle and have poor water resistance, which limit their applications as high performance films or coatings [30]. To improve the properties of soy protein-based films, various modification methods [29], such as blending with plasticizers and polymers (e.g., PUs, polycaprolactone, and poly(butylene succinate)), chemical cross-linking, and enzyme and surface modification, have been tested, with blending being the most useful method to improve film properties.

Two methods have been successfully developed to improve the compatibility between protein and synthetic PUs during the production of protein/PU blend films. One is to blend modified oil-soluble protein products with hydrophobic PUs; the other is to blend protein feedstocks with waterborne PUs. In one study, a hydrophobic soy protein derivative (PDSP, p-phenylene diamine soy protein) synthesized via the modification of SPI with p-phenylene diamine was blended with castor oil-based PU in a mixed solvent of N, N-dimethyl formamide (DMF) and dimethyl sulfoxide (DMSO) and then used to produce a series of blend films by casting [31]. Relatively good compatibility and strong hydrogen-bond interactions between castor oil-based PU and PDSP components were observed. With increasing PU content, the blend films showed decreased tensile strength and Young's modulus, while the elongation at break increased and the thermal stability and water resistance improved [31]. Waterborne polyurethanes (WPUs) are promising alternatives, which have environmentally friendly features compared to solvent-borne PUs, and have attracted much attention for the preparation of protein/PU blend films or coatings [32–36]. Blend films from an SPI alkaline solution and polypropylene glycol (PPG)-based anionic WPU exhibited high transparency and good compatibility as a result of strong hydrogen-bond interactions [32]. SPI/PPG-based WPU blend films also showed decreased tensile strength (from 9 to 3 MPa), decreased Young's modulus (from 110 to 10 MPa), increased elongation at break (from 71 to 365 %), and improved water resistance as the WPU content increased from 0 to 50 % [32]. The blend films had lower cytotoxicity than WPU films, which might be attributed to the hydrolysis of soy protein which provided nutrition for the cell cultured in the media [32]. A potential application as a biomedical material was suggested for the SPI/PPG-based WPU blend films. Blend films from poly(butylene adipate) (PBA)-based anionic WPU and SPI and from poly(ε-caprolactone) (PCL) glycol-based cationic WPU and wheat gluten also showed good compatibility and improved water resistance and mechanical properties [33, 34]. Through blending of a thermally polymerized whey protein isolate (PWPI) and a commercial PU dispersion, wood finish coatings with good mold resistance were synthesized [35].

4.4.3   Composites/Plastics

PU composites/plastics based on proteins have been developed by modification with PU prepolymers or isocyanates [37–39]. For example, a series of PU composites from soy dreg, soy flour, or SPI with a 30–50 % addition of a PPG-based PU prepolymer were prepared by a compression-molding process [37]. With increasing content of the PU prepolymer, the composite materials showed more homogeneous cross-section and improved compatibility due to more chemical bonding via reactions between isocyanate groups in the PU prepolymer and active groups in three soy protein products. Depending on the type of soy protein product and content of the PU prepolymer, the resulting composites showed properties that varied from plastic to elastomer [37]. For all three soy protein products, addition of the PU prepolymer improved water resistance of the PU composites. Another type of soy dreg-based PU composite was prepared by using a castor oil-based PU prepolymer via the same compression-molding process [38]. As the molar ratio of isocyanate to hydroxyl groups increased from 1.33 to 2.0 during the preparation of the castor oil-based PU prepolymer, the PU composites exhibited increased tensile strength and improved water and organic solvent resistance. This may have been due to the increased degree of crosslinking, which weakened the penetration and swelling capabilities of water and organic solvents such as toluene [38].

Zein-based PUs were synthesized with several isocyanate compounds, such as phenyl isocyanate (PI), n-hexyl isocyanate (HI), isophorone diisocyanate (IPDI), and methylenediphenyl 4,4′-diisocyanate (MDI) [39]. Nuclear magnetic resonance (NMR) analysis and model compounds were used to identify major reaction pathways that occurred between zein and isocyanates. Through modification, zein-based PUs were more resistant to water and showed varied mechanical properties with changes in the isocyanate type and content [39]. Potential applications of zein-based PUs in bioplastics, floor coatings, and pigment coatings for paper have been suggested.

4.5   Biodegradability of Protein-Based Polyurethane Products

Biodegradation is the decomposition of materials as a result of the action of naturally-occurring microorganisms such as bacteria and fungi [40]. Biodegradability is desirable for materials with single-use or short-term applications, such as packaging materials, but are unfavorable for materials with long-term applications, such as engineering materials for automotive and construction uses [41]. Studies have been conducted on the biodegradability of protein-based PU products, such as foams, films, and composites. Increased degradation of flexible PU foams during soil incubation tests was observed as the SPI fraction increased in the SPI/polyether polyol blend [42]. In the initial 14 days, all SPI-based PU foams showed relatively fast degradation rates. This mainly resulted from the degradation of non-crosslinked SPI [42]. Approximately 12.6 % of liquefied DDG-derived PU foams were degraded after 10 months of soil incubation [24]. Composite materials from soy dreg and 40 % castor oil-based PU prepolymer had approximately 35 % degradation after an 87-day soil incubation [38]. With an increase in the molar ratio of isocyanate to hydroxyl groups from 1.33 to 2.0 during the preparation of the castor oil-based PU prepolymer, the degradation of the resulting soy dreg-based PU composite decreased because of an increase in the degree of crosslinking. Scanning electron microscopy (SEM) revealed an increase in the number and size of holes on the surface of the biodegraded composites compared to the original surface [38]. After a 15-day soil incubation, 61.9 % weight loss was observed for the thermal compression-molded films made from a blend of wheat gluten and PCL glycol-based cationic WPU at a weight ratio of 5:1 [34]. The high biodegradation degree of protein/WPU blend films was due to the absence of chemical crosslinking between wheat gluten and PU components and the biodegradability of PCL glycol and wheat gluten [34].

4.6   Summary and Future Aspects

Protein-based feedstocks from soy, corn, wheat, and algae have been used for the production of PU products such as foams, films, coatings, and composites. There is significant potential for their wide applications in the PU industry. Currently, blending of protein-based feedstocks with petroleum-based polyols or other polymers has been extensively studied for PU applications. Compared to petroleum-based PU products, protein-based analogues exhibited comparable or improved performance. Recently, there has been increasing interest in exploring other approaches for the production of polyols and PUs from protein-based feedstocks, including liquefaction, hydrolysis followed by oxypropylation, and hydrolysis followed by reactions with ethylene diamine and then ethylene carbonate. It is believed that the development of new synthetic methods for effective modification of many functional groups in proteins will be a promising approach to produce value-added polymeric materials from protein-based feedstocks for more extensive applications.

References

1. Schulz GE, Schirmer RH (1979) Principles of protein structure. Springer, Berlin
2. Hettiarachchy N, Kalapathy U (1997) Soybean protein products. In: Liu K (ed) Soybeans: chemistry, technology, and utilization. International Thomson Publishing, Singapore
3. Wool R, Sun XS (2011) Bio-based polymers and composites. Elsevier, London
4. Mu Y, Wan X, Han Z, Peng Y, Zhong S (2012) Rigid polyurethane foams based on activated soybean meal. J Appl Polym Sci 124:4331–4338
5. Kumar R, Choudhary V, Mishra S, Varma IK, Mattiason B (2002) Adhesives and plastics based on soy protein products. Ind Crop Prod 16:155–172
6. Kinsella JE (1979) Functional properties of soy proteins. J Am Oil Chem Soc 56:242–258
7. Zhang L, Chen P, Huang J, Yang G, Zheng L (2003) Ways of strengthening biodegradable soy-dreg plastics. J Appl Polym Sci 88:422–427
8. Lim TK (2012) Glycine max. Edible medicinal and non-medicinal plants. Springer, Netherlands, pp 634–714
9. Shukla R, Cheryan M (2001) Zein: the industrial protein from corn. Ind Crop Prod 13:171–192
10. Anderson TJ, Lamsal BP (2011) Zein extraction from corn, corn products, and coproducts and modifications for various applications: a review. Cereal Chem 88:159–173
11. Kuktaite R, Türe H, Hedenqvist MS, Gällstedt M, Plivelic TS (2014) The gluten biopolymer and nano-clay derived structures in wheat gluten-urea-clay composites: relation to barrier and mechanical properties. ACS Sustain Chem Eng 2:1439–1445
12. Wieser H (2007) Chemistry of gluten proteins. Food Microbiol 24:115–119
13. Xu J, Bietz JA, Carriere CJ (2007) Viscoelastic properties of wheat gliadin and glutenin suspensions. Food Chem 101:1025–1030
14. Domenek S, Feuilloley P, Gratraud J, Morel MH, Guilbert S (2004) Biodegradability of wheat gluten based bioplastics. Chemosphere 54:551–559
15. Chisti Y (2007) Biodiesel from microalgae. Biotech Adv 25:294–306
16. Molina Grima E, Belarbi EH, Acién Fernández FG, Robles Medina A, Chisti Y (2003) Recovery of microalgal biomass and metabolites: process options and economics. Biotech Adv 20:491–515
17. Satyanarayana KG, Mariano AB, Vargas JVC (2011) A review on microalgae, a versatile source for sustainable energy and materials. Int J Energ Res 35:291–311
18. Neveux N, Yuen AKL, Jazrawi C, Magnusson M, Haynes BS, Masters AF, Montoya A, Paul NA, Maschmeyer T, de Nys R (2014) Biocrude yield and productivity from the hydrothermal liquefaction of marine and freshwater green macroalgae. Bioresour Technol 155:334–341
19. Li H, Liu Z, Zhang Y, Li B, Lu H, Duan N, Liu M, Zhu Z, Si B (2014) Conversion efficiency and oil quality of low-lipid high-protein and high-lipid low-protein microalgae via hydrothermal liquefaction. Bioresour Technol 154:322–329
20. Park S, Li Y (2012) Evaluation of methane production and macronutrient degradation in the anaerobic co-digestion of algae biomass residue and lipid waste. Bioresour Technol 111:42–48
21. Beckman EJ, Russell AJ (1996) Protein-containing polymers and a method of synthesis of protein-containing polymers in organic solvents. US Patent 5482996
22. Narayan R, Hablot E, Graiver D, Sendijarevic V (2014) Soy meal-based polyols for rigid polyurethane foams. PU Mag 11(3)
23. Kumar S, Hablot E, Moscoso JLG, Obeid W, Hatcher PG, DuQuette BM, Graiver D, Narayan R, Balan V (2014) Polyurethane preparation using proteins obtained from microalgae. J Mater Sci 49:7824–7833
24. Yu F, Le Z, Chen P, Liu Y, Lin X, Ruan R (2008) Atmospheric pressure liquefaction of dried distillers grains (DDG) and making polyurethane foams from liquefied DDG. Appl Biochem Biotechnol 148:235–243

25. Chang LC, Xue Y, Hsieh FH (2001) Comparative study of physical properties of water-blown rigid polyurethane foams extended with commercial soy flours. J Appl Polym Sci 80:10–19
26. Lin Y, Hsieh F, Huff HE, Iannotti E (1996) Physical, mechanical, and thermal properties of water-blown rigid polyurethane foam containing soy protein isolate. Cereal Chem 73:189–196
27. Lin Y, Hsieh F, Huff HE (1997) Water-blown flexible polyurethane foam extended with biomass materials. J Appl Polym Sci 65:695–703
28. Park SK, Hettiarachchy NS (1999) Physical and mechanical properties of soy protein-based plastic foams. J Am Oil Chem Soc 76:1201–1205
29. Song F, Tang DL, Wang XL, Wang YZ (2011) Biodegradable soy protein isolate-based materials: a review. Biomacromolecules 12:3369–3380
30. Kumar R, Liu D, Zhang L (2008) Advances in proteinous biomaterials. J Biobased Mater Bioenergy 2:1–24
31. Liu D, Tian H, Zhang L, Chang PR (2008) Structure and properties of blend films prepared from castor oil-based polyurethane/soy protein derivative. Ind Eng Chem Res 47:9330–9336
32. Tian H, Wang Y, Zhang L, Quan C, Zhang X (2010) Improved flexibility and water resistance of soy protein thermoplastics containing waterborne polyurethane. Ind Crop Prod 32:13–20
33. Zhang M, Song F, Wang XL, Wang YZ (2012) Development of soy protein isolate/waterborne polyurethane blend films with improved properties. Colloids Surf B Biointerfaces 100:16–21
34. Zhong N, Yuan Q (2013) Preparation and properties of molded blends of wheat gluten and cationic water-borne polyurethanes. J Appl Polym Sci 128:460–469
35. Wright NC, Li J, Guo M (2006) Microstructural and mold resistant properties of environment-friendly oil-modified polyurethane based wood-finish products containing polymerized whey proteins. J Appl Polym Sci 100:3519–3530
36. Madbouly SA, Lendlein A (2012) Degradable polyurethane/soy protein shape-memory polymer bBlends prepared via environmentally-friendly aqueous dispersions. Macromol Mater Eng 297:1213–1224
37. Chen Y, Zhang L, Du L (2003) Structure and properties of composites compression-molded from polyurethane prepolymer and various soy products. Ind Eng Chem Res 42:6786–6794
38. Chen Y, Zhang L, Deng R, Cui Y (2007) A new network composite material based on soy dreg modified with polyurethane prepolymer. Macromol Mater Eng 292:484–494
39. Sessa DJ, Cheng HN, Kim S, Selling GW, Biswas A (2013) Zein-based polymers formed by modifications with isocyanates. Ind Crop Prod 43:106–113
40. Gómez EF, Luo X, Li C, Michel FC Jr, Li Y (2014) Biodegradability of crude glycerol based polyurethane foams during composting, anaerobic digestion and soil incubation. Polym Degrad Stab 102:195–203
41. Petrović ZS (2008) Polyurethanes from vegetable oils. Polym Rev 48:109–155
42. Wang G, Zhou A (2011) Soy protein based biodegradable flexible polyurethane foam. Adv Mater Res 152:1862–1865

Made in the USA
Monee, IL
21 November 2023

47051025R00057